Lattice-Ordered Groups

An Introduction

by

Marlow Anderson
The Colorado College, Colorado Springs,
Colorado, U.S.A.

and

Todd Feil
Department of Mathematical Sciences, Denison University,
Granville, Ohio, U.S.A.

D. Reidel Publishing Company
A MEMBER OF THE KLUWER ACADEMIC PUBLISHERS GROUP
Dordrecht / Boston / Lancaster / Tokyo

Library of Congress Cataloging in Publication Data

CIP

Anderson, Marlow, 1950–
Lattice-ordered groups : an introduction / by Marlow Anderson and Todd Feil.
p. cm.—(Reidel texts in the mathematical sciences)
Bibliography: p.
Includes indexes.
ISBN-13: 978-94-010-7792-7 e-ISBN-13: 978-94-009-2871-8
DOI: 10.1007/ 978-94-009-2871-8
1. Lattice ordered groups. I. Feil, Todd, 1951– II. Title.
III. Series.
QA171.A553 1987
512′.22—dc 19 87–28598
CIP

Published by D. Reidel Publishing Company,
P.O. Box 17, 3300 AA Dordrecht, Holland.

Sold and distributed in the U.S.A. and Canada
by Kluwer Academic Publishers,
101 Philip Drive, Norwell, MA 02061, U.S.A.

In all other countries, sold and distributed
by Kluwer Academic Publishers Group,
P.O. Box 322, 3300 AH Dordrecht, Holland.

Lattice-Ordered Groups

An Introduction

Reidel Texts in the Mathematical Sciences

A Graduate-Level Book Series

Preface

The study of groups equipped with a compatible lattice order ("lattice-ordered groups" or "ℓ-groups") has arisen in a number of different contexts. Examples of this include the study of ideals and divisibility, dating back to the work of Dedekind and continued by Krull; the pioneering work of Hahn on totally ordered abelian groups; and the work of Kantorovich and other analysts on partially ordered function spaces.

After the Second World War, the theory of lattice-ordered groups became a subject of study in its own right, following the publication of fundamental papers by Birkhoff, Nakano and Lorenzen. The theory blossomed under the leadership of Paul Conrad, whose important papers in the 1960s provided the tools for describing the structure for many classes of ℓ-groups in terms of their convex ℓ-subgroups. A particularly significant success of this approach was the generalization of Hahn's embedding theorem to the case of abelian lattice-ordered groups, work done with his students John Harvey and Charles Holland. The results of this period are summarized in Conrad's "blue notes" [C].

Holland's proof in 1963 that every lattice-ordered group can be represented as a group of order-preserving permutations of a totally ordered set has been of decisive importance in the further development of the theory of nonabelian ℓ-groups, since it is the only tool available for studying arbitrary ℓ-groups. In particular, this approach has been essential in the study of varieties of ℓ-groups, a subject of major interest in the last 15 years, with important contributions being made by Holland and his students Stephen McCleary and Andrew Glass. In fact, Holland's theorem has led to considerable study of the permutation groups of totally ordered sets in their own right. This work is summarized in Glass' book *Ordered Permutation Groups* [G].

The present work intends to focus narrowly on a concise exposition of the classical theory of lattice-ordered groups. We consequently omit most work on totally ordered groups (for which, the reader may see the books of Kopytov and Kokorin [KK] and of Rhemtulla and Mura [MR]). We also do not consider other partially ordered structures, such as partially ordered rings or semigroups; some material on such subjects may be found in Fuchs' book [F]. In particular, we do not discuss lattice-ordered rings; they figure in Bigard, Keimel and Wolfenstein [BKW]. Our emphasis is algebraic rather than analytic; for the vast literature on partially ordered linear spaces, the reader should consult Schaefer [S], Vulich [V], Luxemberg and Zaanen [LZ] or Aliprantis and Burkinshaw [AB]. Finally, we use the techniques of lattice-ordered permutation groups only in so far as is necessary to obtain results in the theory of ℓ-groups; we consequently view Glass' book [G] as an essential but complementary volume to our own.

After a chapter of introductory material on the lattice of convex ℓ-subgroups of a lattice-ordered group, we prove in the next several chapters the four important representation theorems of the theory, in increasing order of generality: the Bernau representation for archimedean ℓ-groups, the Conrad-Harvey-Holland representation for abelian ℓ-groups, the Lorenzen representation, and finally the Holland representation for all ℓ-groups. We next describe free lattice-ordered groups, preparatory to Chapter 7, which is an extended exposition of the theory of varieties of ℓ-groups. Included in this chapter is a proof of Holland's theorem that the class of normal-valued ℓ-groups is the largest proper ℓ-group variety. The theory of ℓ-group completions is the subject of the next two chapters; we include a proof of the existence of a lateral completion for any lattice-ordered group, a result of major difficulty and importance, first proved by Bernau in the early seventies. In the next chapter we describe the theory of finite-valued and special-valued ℓ-groups, with emphasis on the root system of values of a normal-valued ℓ-group, and analogies that can be drawn with the Conrad-Harvey-Holland theorem. The final chapter of the text provides an exposition of the theory of lattice-ordered groups of divisibility of commutative rings. What follows is an appendix, which is devoted to an extended compendium of examples of ℓ-group classes and ℓ-groups themselves, which illustrate the boundaries of the theory. Finally, we provide a comprehensive bibliography of papers in lattice-ordered groups, with special emphasis on the last twenty years.

The text assumes only the usual graduate courses in algebra, analysis and point-set topology, and consequently should be accessible to graduate students and professional mathematicians in other fields; we hope it also offers something to the expert.

The exposition is for the most part self-contained; we do occasionally discuss without proof further developments in the theory. The major exception to this is in Chapter 7, where we quote without proof major theorems from the theory of ℓ-permutation groups; the interested reader may consult the proofs published in Glass' book [G].

Additive notation is used for the group operation into Chapter 5 where it is shown that each ℓ-group can be considered as an ℓ-subgroup of the ℓ-group of ordered preserving permutations of a totally ordered set. From there until the end, with the exceptions of Chapters 8 and 11, multiplicative notation is used. It has become somewhat of a convention that abelian groups be written additively and non-abelian groups multiplicatively. This convention is by no means universal, however. We have tried to be consistent within chapters and believe that in this book, the change over to multiplicative notation makes sense where it occurs. Chapters 8 and 11 use additive notation since Section 8.2 concerns archimedean (and hence abelian) ℓ-groups and Chapter 11 deals with abelian ℓ-groups.

References in the text to examples of ℓ-groups which appear in the appendix are in the form E#; references to discussions in the appendix of ℓ-group classes are by the script letter denoting the class; all such abbreviations appear in the index. The bibliography consists of a list of pertinent books, followed by a list of papers and dissertations; we refer to books by authors' initials [XY], and to papers by the authors' names, followed by two digits indicating the year. The reader should consult the appendix and the bibliography for more detail.

We owe a considerable debt of gratitude to Paul Conrad and Charles Holland, on several grounds. Their mathematical influence can be seen on every page; note particularly that the earlier material in the book is heavily influenced by Conrad's "blue notes" [C]. Their personal examples as our thesis advisors have been inspirational for our careers as mathematicians. Both men have also examined various portions of the manuscript and offered their comments. Charles Holland taught a course from a preliminary draft, and gathered many valuable corrections and comments. The extensive comments from Andrew Glass have improved the manuscript a great deal. We have also benefited from informal discussions of the project with many mathematicians across the country. As for any errors (typographical or otherwise) which remain, we of course take the only reasonable position, and blame one another.

Our respective institutions deserve thanks for the financial support we have received to expedite this project, in the face of our considerable geographic separation. Their aid also enabled us to prepare this manuscript using TEX, which has improved its appearance and readability, even in the hands of novice users.

Finally, we wish to dedicate this book to Audrey and Robin.

Marlow Anderson
The Colorado College

Todd Feil
Denison University

Contents

Chapter 1: *Fundamentals*

Section 1.1: *Preliminaries and Basic Examples*

In this section we shall develop the basic definitions and notations used in the theory of lattice-ordered groups, and provide a set of examples whose significance will be more fully appreciated when we have proved the basic representation theorems for lattice-ordered groups in Chapters 2 through 5.

Additive and multiplicative notation for the underlying group of a lattice-ordered group are both used, depending on the particular application in mind, and we shall feel free to use both. In order that the word "positive" have its customary meaning, we shall in our initial definitions use additive notation. The reader should not infer from this that all groups referred to are abelian.

A group $(G,+)$ is said to be *partially ordered* if it is equipped with a reflexive, antisymmetric and transitive relation $\leq$ (a *partial order*) which is compatible with $+$; that is, if $g \leq h$, then $g+k \leq h+k$ and $k+g \leq k+h$, for any $g,h,k \in G$. The set $G^+ = \{g \in G : g > 0\}$ of positive elements is called the *positive cone of* G. It is easily verified that the partial order is determined by G^+, in the sense that $g \leq h$ if and only if $h-g \in G^+ \cup \{0\}$. Furthermore, any set $P \subseteq G$ which is closed under addition, normal (i.e., $g+P-g=P$ for all $g \in G$), and satisfies the property $P \cap -P = \emptyset$, gives rise to a partial order on G.

Suppose now that the partial order on G is in fact a *lattice* (meaning that each pair of elements a,b of G has a least upper bound (join) $a \vee b$ and a greatest lower bound (meet) $a \wedge b$). Then G is a *lattice-ordered group* (*ℓ-group*) (some authors prefer the term lattice group). In cases where we wish to speak of joins (or meets) of sets with cardinality greater than two, we shall use notation such as the following: $a_1 \vee a_2 \vee \ldots a_n$; $\bigvee_{i\in I} a_i$ (for some index set I); $\vee A$ (for some set $A \subseteq G$); or the analogues for meets. Notice, however, that the definition of ℓ-group does not necessarily imply the existence of joins and meets for other than finite sets.

If the lattice order on an ℓ-group G is a total order (that is, for any $g,h \in G$, either $g \geq h$ or $h \geq g$), we then call G an *o-group*. Notice that in this case the positive cone satisfies the additional criterion that $P \cup -P \cup \{0\} = G$. Just as a semigroup theorist considers his problem done when it is reduced to group theory, we shall adopt the view that our understanding is sufficient when we can reduce ℓ-group questions to questions regarding o-groups. The subject of o-groups is itself a rich and difficult one, and the reader is advised to consult [MR] or [KK] for an introduction to it.

It is easy to verify that the group operation distributes over both meet and join; that is,

$$a+b\vee c+d=(a+b+d)\vee(a+c+d),$$

and likewise for $\wedge$. The relationship between the lattice operations and the group inverse is given by what might be called DeMorgan's laws:

Proposition 1.1.1 Let $a, b \in G$. Then:

(a) $-(a \vee b) = -a \wedge -b$.

(b) $-(a \wedge b) = -a \vee -b$.

Proof: We shall prove only (a). Since $a, b \leq a \vee b$ then $-a, -b \geq -(a \vee b)$, and so $-a \wedge -b \geq -(a \vee b)$. For the reverse inequality, suppose that $-a, -b \geq c$. Then, $a, b \leq -c$, and so $a \vee b \leq -c$; that is $c \leq -(a \vee b)$.//

We may infer from the proof of this proposition a further result. Namely, a partially ordered group which is a join-semilattice (or meet-semilattice) is necessarily an ℓ-group. A useful identity (due to Birkhoff [42]) also follows from the proposition:

1.1.2: $a - (a \wedge b) + b = a \vee b$;

this holds because $a - (a \wedge b) = 0 \vee (a - b) = (a \vee b) - b$.

For $g \in G$, the *positive part* g^+ of g is $g \vee 0$; the *negative part* g^- is $(-g) \vee 0$. Then

$$g + g^- = g + (-g \vee 0) = 0 \vee g = g^+ ;$$

and so we have that $g = g^+ - g^-$. Now,

$$g^+ \wedge g^- = (g + g^-) \wedge g^- = (g \wedge 0) + g^- = -g^- + g^- = 0.$$

If $a \wedge b = 0$, we say that a and b are *disjoint*.

Thus, we have shown that each element of an ℓ-group can be represented as the difference of two disjoint positive elements; the reader may verify that this representation is unique. In fact, this property characterizes ℓ-groups: a directed partially ordered group in which each element may be uniquely represented as the difference of disjoint positive elements is necessarily an ℓ-group (see [F]).

Note that disjoint elements commute; in fact if $a \wedge b = 0$ then $a + b = a \vee b$. For,

$$(a + b) - (a \vee b) = (a + b) + (-a \wedge -b) = (a + b - a) \wedge a = a + (b \wedge a) - a = 0.$$

The *absolute value* $|g|$ of g is $g^+ + g^-$; like ordinary absolute value, $g = |g|$ exactly when $g \geq 0$. The reader may verify that $|g| = g^+ \vee g^- = g \vee -g$. The absolute value satisfies a weakened triangle inequality:

1.1.3a: $|g + h| \leq |g| + |h| + |g|$.

For,

$$|g + h| = (g + h) \vee 0 + (-h - g) \vee 0 \leq (g \vee 0) + (h \vee 0) + (-h \vee 0) + (-g \vee 0) \leq |g| + |h| + |g|.$$

Furthermore, since $(g + h)^+$ and $(g + h)^-$ are disjoint and so commute, it also follows that:

1.1.3b: $|g + h| \leq |h| + |g| + |h|$.

Note that a slight modification of the proof above shows that if the ℓ-group is abelian, the usual triangle inequality holds. In fact, the converse holds: any ℓ-group satisfying the usual triangle inequality is necessarily abelian (Kalman [60]).

An important law which ℓ-groups obey is the *Reisz decomposition property*, which is the content of the next proposition. There exist partially ordered groups which have this property, which are not ℓ-groups; they have been much studied (see for example Fuchs [65] or Loy and Miller [72]).

Proposition 1.1.4 Suppose $h_1, \ldots, h_n \in G^+$. If $0 \leq g \leq h_1 + h_2 + \ldots + h_n$, then $g = g_1 + g_2 + \ldots g_n$, where $0 \leq g_i \leq h_i$.

Proof: We proceed by induction on n. The result is clear for $n = 1$. Suppose now that $0 \leq g \leq h_1 + h_2 + \ldots h_n$. Let $g_1 = g \wedge h_1$. Then

$$k = -g_1 + g = (-g \vee -h_1) + g = 0 \vee (-h_1 + g) \leq h_2 + \ldots + h_n,$$

and so by induction $k = g_2 + \ldots + g_n$, with $0 \leq g_i \leq h_i$. But since $g = g_1 + k$ the desired conclusion follows. //

Note the following consequence of the Reisz decomposition property:

1.1.5 for positive g,h and k , $g \wedge (h + k) \leq (g \wedge h) + (g \wedge k)$.

It is natural to inquire as to which groups and which lattices admit ℓ-group structure. Some of the more obvious restrictions are contained in the following proposition. We require the following terminology from lattice theory (see [B]): A lattice L is *homogeneous* if for $a, b \in L$, there exists a lattice automorphism of L taking a to b; L is *distributive* if $a \vee (b \wedge c) = (a \vee b) \wedge (a \vee c)$, and dually, for all $a, b, c \in L$.

Proposition 1.1.6

(a) The lattice of an ℓ-group is homogeneous and distributive.
(b) The group of an ℓ-group is torsion free.
(c) In an ℓ-group, $na = nb$ implies $-c + a + c = b$ for some c.

Proof: (a): The homogeneity of the lattice follows from the fact that $x \mapsto x + g$ is a lattice automorphism, for any fixed $g \in G$. To show that the lattice is distributive, we need only show that if $g \wedge k = h \wedge k$ and $g \vee k = h \vee k$, then $g = h$(see [B]). But because of 1.1.2, we have

$$g = (g \vee k) - k - (g \wedge k) = (h \vee k) - k - (h \wedge k) = h.$$

(b): Suppose that $ng = 0$, for some ℓ-group element g and positive integer n. Then (by repeatedly distributing addition over joins),

$$n(g \vee 0) = ng \vee (n-1)g \vee \ldots \vee g \vee 0 = (n-1)g \vee \ldots \vee g \vee 0 = (n-1)(g \vee 0).$$

But then, $g \vee 0 = 0$. Similarly, $-g \vee 0 = 0$, and so $g = 0$.

(c): Let $c = (n-1)a \vee (n-2)a + b \vee \cdots \vee a + (n-2)b \vee (n-1)b$. Then $a + c = c + b$. //

That the lattice of an ℓ-group is distributive was known in the abelian case by Dedekind [97] and rediscovered by Freudenthal [36]. Birkhoff [42] proved the general theorem.

The conditions of Proposition 1.1.6 do not, however, characterize those lattices or those groups which admit ℓ-group structure (see E2 and E10). Indeed, there are presently no satisfactory answers to the "ℓ-groupability" question for either lattices or groups; for further discussion, see Conrad [67b] and Berman [74]. However, there is a nice characterization of subgroups of lattice-orderable groups which we will discuss in Chapter 5.

The condition of torsion-freeness does imply lattice-orderability in the abelian case (Levi [13]):

Proposition 1.1.7 For an abelian group G, the following are equivalent:

(a) G admits an o-group structure.
(b) G admits an ℓ-group structure.
(c) G is torsion free.

Proof: (a)⇒(b) is obvious, and (b)⇒(c) is contained in Proposition 1.1.6. If G is torsion free and abelian, it may be embedded into its divisible hull G^d, which as a rational vector space may be represented as $\sum_{\lambda\in\Lambda} \mathbf{Q}_\lambda$, a direct sum of copies of the additive rationals $\mathbf{Q}$. Provide Λ with a well-ordering, and call an element in G^d positive if it is positive in its first nonzero component according to the well-ordering of Λ. It can be checked that this provides a total order for G^d, and consequently for G.//

Proposition 1.1.8 Let g, h be elements of the ℓ-group G. Then

$$g^+ + h^+ \geq (g+h)^+ \geq g^+ \wedge h^+.$$

Proof: The first inequality is easily proved as follows:

$$g^+ + h^+ = (g \vee 0) + (h \vee 0) = (g+h) \vee g \vee h \vee 0 \geq (g+h) \vee 0 = (g+h)^+.$$

For the proof of the second inequality, we have

$$\begin{aligned}(g+h)^+ &= (g+h) \vee 0 = (g \vee -h) + h = (g \vee -h)^+ - (g \vee -h)^- + h\\ &= (g \vee -h \vee 0) - ((-g \wedge h) \vee 0) + h\\ &= g^+ \vee h^- - ((-g \vee 0) \wedge (h \vee 0)) + h\\ &= (g^+ \vee h^-) - (g^- \wedge h^+) + h.\end{aligned}$$

Now apply 1.1.2 to the join in this last expression to obtain

$$(g+h)^+ = g^+ - (g^+ \wedge h^-) + h^- - (g^- \wedge h^+) - h^- + h^+.$$

Since $g^- \wedge h^+$ and h^- are disjoint and hence commute, we have

$$\begin{aligned}(g+h)^+ &= g^+ - (g^+ \wedge h^-) - (g^- \wedge h^+) + h^+\\ &= g^+ - ((g^- \wedge h^+) + (g^+ \wedge h^-)) + h^+\\ &= g^+ - ((g^- \wedge h^+) \vee (g^+ \wedge h^-)) + h^+,\end{aligned}$$

where this last equality holds because $g^- \wedge h^+$ and $g^+ \wedge h^-$ are disjoint. But then

$$(g+h)^+ \geq g^+ - (h^+ \vee g^+) + h^+ = g^+ \wedge h^+$$

(the last equality being 1.1.2).//

We shall now describe some important equalities which involve distributing operations over infinite meets and joins. Care must be taken in interpreting these equations, since not all infinite meets and joins need exist in an arbitrary ℓ-group.

First, we mention the infinite versions of DeMorgan's laws, namely

1.1.9: $-(\bigvee h_\lambda) = \bigwedge(-h_\lambda)$, and dually.

By these equations we mean that if the infinite lattice operation on one side exists, then so does the one on the other side, and equality holds. The proof of these laws is nearly identical to that of the finite case 1.1.1, as the reader may verify.

Next, we show that the group operation distributes over infinite lattice operations; we interpret the existence of the infinite lattice operations in the same way that we did for 1.1.9:

1.1.10: $g + \bigvee h_\lambda = \bigvee(g + h_\lambda)$, and dually.

For an example of the proofs necessary, suppose that $\bigvee h_\lambda$ exists. It is evident that $g+\bigvee h_\lambda \geq g+h_\lambda$, for all λ . Suppose that $k \geq g+h_\lambda$ for all λ; we then have that $-g+k \geq h_\lambda$, and so $-g+k \geq \bigvee h_\lambda$. We thus have $k \geq g + \bigvee h_\lambda$, and hence $g + \bigvee h_\lambda = \bigvee(g+h_\lambda)$, as claimed.

Furthermore, we show that meets distribute over infinite joins, and dually (note that this is by no means the case in an arbitrary distributive lattice). That is, we have the following:

1.1.11: if $\bigvee h_\lambda$, exists, then $g \wedge \bigvee h_\lambda = \bigvee(g \wedge h_\lambda)$, and dually.

Note that the existence of the infinite lattice operation on the right need not imply existence on the left (for example, consider a disjoint set of elements in a cardinal sum of integers, defined in 1.1.14). For a proof, suppose that $s = \bigvee h_\lambda$ exists. Then $g \wedge h_\lambda \leq g \wedge s$, for all λ , and so $0 \leq (g \wedge s) - (g \wedge h_\lambda) \leq g \wedge (s - h_\lambda) \leq s - h_\lambda$. But $0 = s - \bigvee h_\lambda = \bigwedge(s-h_\lambda)$, and so $0 = \bigwedge((g \wedge s) - (g \wedge h_\lambda)) = g \wedge s - \bigvee(g \wedge h_\lambda)$, as required.

We shall now examine what can be said about an expression involving perhaps infinitely many meets and joins. We say that an ℓ-group *satisfies the generalized distributive laws* if

$$\bigvee_I \bigwedge_J g_{ij} = \bigwedge_{J^I} \bigvee_I g_{if(i)}, \text{ and dually for all index sets } I \text{ and } J.$$

A tedious inductive application of the elementary distributive laws shows that this is certainly the case for any distributive lattice, if I and J are finite. Furthermore, condition 1.1.11 gives us that these laws hold for an arbitrary ℓ-group if I is finite. Unfortunately, the generalized distributive laws need not hold for all ℓ-groups (see E19); ℓ-groups which do satisfy them are called *completely distributive*. We will characterize completely distributive ℓ-groups in Chapter 10.

We shall conclude this section with a list of examples of ℓ-groups. The representation theorems of Chapters 2 through 5 will reveal the inclusiveness of this list.

1.1.12. Let X be a topological space and $C(X)$ the additive group of real-valued continuous functions. We make $C(X)$ an ℓ-group by providing it with its usual pointwise order: $f \leq g$ if and only if $f(x) \leq g(x)$, for all $x \in X$. Obvious modifications of this basic example abound in analysis.

1.1.13. Let Γ be a *root system*. That is, Γ is a partially ordered set for which $\{\alpha : \alpha \geq \gamma\}$ is totally ordered, for any $\gamma \in \Gamma$. Let $\{H_\gamma : \gamma \in \Gamma\}$ be a collection of o-groups indexed by Γ. Consider functions v on Γ for which $v(\gamma) \in H_\gamma$, for all $\gamma \in \Gamma$. Given such a function v, the subset of Γ where v is not zero is called its *support* and is denoted by $spt(v)$. Let $V(\Gamma, H_\gamma)$ be the set of all such functions whose support satisfies the ascending chain condition. It is easy to see that this is a group under addition. Furthermore, if we define an element to be positive if it is positive at each maximal element of its support, the reader may verify that $V(\Gamma, H_\gamma)$ is an ℓ-group, which we shall call a *Hahn group* on Γ. Now consider those elements of $V(\Gamma, H_\gamma)$ whose supports are finite. It is evident that this is also an ℓ-group when equipped with the same order, which we denote by $\Sigma(\Gamma, H_\gamma)$, and call a *restricted* Hahn group. The case where each H is the o-group $\mathbf{R}$ of real numbers is of particular importance; Hahn [07] made use of such groups for Γ totally ordered in his pioneering work on abelian o-groups.

1.1.14. Let $\{H_\lambda : \lambda \in \Lambda\}$ be a collection of o-groups. We denote by ΣH_λ and ΠH_λ the restricted, and unrestricted, direct products of the H's, respectively. Both of these groups can be made into ℓ-groups with the *cardinal* (or *pointwise*) order; that is, an element is positive exactly if it is positive at each component. Notice that the restricted and unrestricted direct products are just the corresponding Hahn groups on the index set Λ , equipped with the trivial order. Of course, other subdirect products of the H's may well be ℓ-groups under this order. Now suppose that Λ is totally ordered. We can then make ΣH_λ an o-group by declaring an element positive if it is positive at the largest (or alternatively, the smallest) element of its support (which is in this case finite); this is just the restricted Hahn group on the totally ordered set Λ . If Λ is well-ordered we can impose a similar order on the entire direct product, thus obtaining an unrestricted Hahn group. All such orderings are called *lexicographic*.

1.1.15. Let T be a totally ordered set, and let $A(T)$ be the set of order-preserving, one-to-one, onto maps from T to T (the *order-preserving permutations* of T). Then $A(T)$ is a group under composition. Call an element α positive if $t\alpha \geq t$, for all $t \in T$. Then $A(T)$ is an ℓ-group.

Section 1.2: *Subobjects and Morphisms*

A complication for the algebra of lattice-ordered groups is that even in the abelian case, most subobjects cannot serve as kernels of morphisms. As we shall see, the practical significance of this for the theory of ℓ-groups is that most attention is paid to subobjects which are not only sublattices and subgroups, but also convex. Fortunately, the lattice of convex ℓ-subgroups of a given ℓ-group contains most of the information one might want about the ℓ-group. (For further explication of this point of view, see Anderson, Conrad and Martinez [86].)

We shall begin with morphisms. A function $\phi : G \to H$ between ℓ-groups G and H is an *ℓ-homomorphism* if it is both a lattice and a group homomorphism. The reader may verify that it is sufficient to check that ϕ is a group homomorphism, and that $\phi(g \vee 0) = \phi(g) \vee 0$, for all $g \in G$. Also, it is clear that if ϕ is an ℓ-homomorphism, then $\phi(G)$ is both a sublattice and a subgroup of H; that is, $\phi(G)$ is an *ℓ-subgroup*.

We shall now characterize the kernels of ℓ-homomorphisms. We first need some definitions. An ℓ-subgroup K of an ℓ-group G is *convex* if $h, k \in K$ and $h < g < k$ imply that $g \in K$. (Because of the rather different meaning of "convex" in analysis, convex ℓ-subgroups are sometimes called *solid* subgroups). The reader may verify that in order to check whether a subgroup K of G is a convex ℓ-subgroup, it is necessary only to verify that if $k \in K, g \in G$, and $|g| \leq |k|$, then $g \in K$. A normal convex ℓ-subgroup is called an *ℓ-ideal.* The next theorem asserts that ℓ-ideals are exactly the kernels of ℓ-homomorphisms, and that the group of cosets of an ℓ-ideal can be ordered in such a way as to make an ℓ-group.

Theorem 1.2.1 Let ϕ be an ℓ-homomorphism from ℓ-group G onto ℓ-group H.

(a) The kernel $Ker(\phi)$ of ϕ is an ℓ-ideal.

(b) If $N$is an ℓ-ideal of G, then the set of right cosets G/N can be provided with an order which makes it an ℓ-group, so that the natural map $\nu : G \to G/N$ is an ℓ-homomorphism.

(c) $G/Ker(\phi)$ is ℓ-isomorphic to H.

Proof: (a) is obvious.

(b): From elementary group theory, we know that G/N is a group, and $\nu : G \to G/N$ is a group homomorphism. Define $N+g \geq N+h$ to mean that there exists $k \in N$ such that $k+g \geq h$. It is easily checked that this definition is independent of coset representatives, and that the resulting relation is reflexive and transitive. To show antisymmetry, suppose that $N+g \geq N+h$ and $N+h \geq N+g$. Then, there exist $r, s \in N$ with $r+g \geq h$ and $s+h \geq g$. Thus, $s+r+g \geq s+h \geq g$, and so $s+r \geq s+h-g \geq 0$. Since N is convex, $h-g \in N$ and so $N+g = N+h$. Thus, we have a partial order on G/N. We now claim that $N+g \vee h = (N+g) \vee (N+h)$. Clearly, $N+g \vee h \geq N+g, N+h$. If $N+d \geq N+g, N+h$, then $a+d \geq g$ and $b+d \geq h$, for $a, b \in N$. But then, $(a \vee b)+d = (a+d) \vee (b+d) \geq g \vee h$, and since N is a sublattice, $N+d \geq N+g \vee h$. Thus, G/N is a join-semilattice. Since G/N is also a group, it is an ℓ-group.

(c): This is now obvious. //

It is worth noting that the proof above verifies that the set of right cosets G/N is in fact a lattice under the order of (b) even if N is not normal.

We shall now examine the set $\mathcal{C}(G)$ of convex ℓ-subgroups of a given ℓ-group G, in some detail. We need a little terminology from lattice theory. A lattice is *complete* if each subset has a least upper bound and greatest lower bound. A complete lattice is *Brouwerian* if it satisfies the identity

$$a \wedge \bigvee_{\Lambda} b_\lambda = \bigvee_{\Lambda} (a \wedge b_\lambda).$$

(This is not the usual definition, but is equivalent to it; see [B].)

The following theorem was proved for the set of ℓ-ideals by Birkhoff [42] and extended to $\mathcal{C}(G)$ by Lorenz [62].

Theorem 1.2.2 Let G be an ℓ-group. Then $\mathcal{C}(G)$ is a complete Brouwerian lattice, which is a sublattice of the lattice of all subgroups of G; that is, the join of an arbitrary collection of convex ℓ-subgroups is the group generated by these subgroups.

Proof: It is apparent that the intersection of any collection of convex ℓ-subgroups of G remains a convex ℓ-subgroup. It follows immediately that $\mathcal{C}(G)$ is a complete lattice, for the join of a collection of convex ℓ-subgroups $\{C_\lambda\}$ is just the intersection of all convex ℓ-subgroups D for which $D \supseteq C_\lambda$.

To prove the last statement of the theorem, suppose that $\{C_\lambda\}$ is a collection of convex ℓ-subgroups, and that C is the group generated by them. In order to see that $C \in \mathcal{C}(G)$, suppose that $c \in C$, $g \in G$ and $|g| \leq |c|$. Now, $c = c_1 + c_2 + \ldots + c_n$, where each $c_i \in C_{\lambda_i}$, some $\lambda_i \in \Lambda$. Then by 1.1.3,

$$g^+ \vee g^- = |g| \leq |c_1 + c_2 + \ldots + c_n| \leq |c_1| + |c_2| + \ldots + |c_{n-1}| + |c_n| + |c_{n-1}| + \ldots + |c_1|,$$

and so, by the Reisz decomposition property (1.1.4), g^+ and g^- can be expressed as sums of elements of the C_λ's , and so $g \in C$.

It remains to show that $\mathcal{C}(G)$ is Brouwerian. For convex ℓ-subgroups B,C_λ , it is clear that $\bigvee_\Lambda (B \cap C_\lambda) \subseteq B \cap \bigvee_\Lambda C_\lambda$. If $g \in B^+ \cap C_\lambda$, then it can be expressed as a sum of positive elements from various C_λ's . But since each such summand is bounded by g, each is an element of B as well. Thus $g \in \bigvee_\Lambda (B \cap C_\lambda)$. //

An element x of a lattice is *compact* if $x \leq \bigvee_\Lambda x_\lambda$ implies that $x \leq \bigvee_F x_\lambda$, for some finite subset F; a lattice is *algebraic* if every element is the join of compact elements. Let $G(g)$ be the smallest convex ℓ-subgroup of ℓ-group G which contains g. The subgroups of the form $G(g)$ are called the *principal convex ℓ-subgroups* of G. It is obvious that each such convex ℓ-subgroup is compact in $\mathcal{C}(G)$,and so $\mathcal{C}(G)$ is algebraic. To show that each compact element of $\mathcal{C}(G)$ is of the form $G(g)$, we shall describe such convex ℓ-subgroups explicitly in the following Proposition:

Proposition 1.2.3 Let $g \in G$. Then $G(g) = \{h \in G : |h| \leq n|g|$, for positive integer $n\}$. Consequently, $G(g) = G(|g|)$. If $g, k > 0$, then $G(g \vee k) = G(g) \vee G(k)$ and $G(g \wedge k) = G(g) \cap G(k)$. Furthermore, any compact element of $\mathcal{C}(G)$ is of the form $G(g)$.

Proof: It is obvious that $\{h \in G : |h| \leq n|g|$, for positive integer $n\}$ contains $G(g)$, and an argument similar to that contained in the proof of Theorem 1.2.2 proves that it is a group, and hence a convex ℓ-subgroup; thus, it equals $G(g)$.

Since $g, k \in G(g \vee k)$, it is obvious that $G(g \vee k) = G(g) \vee G(k)$. Also, since $g \wedge k \in G(g) \cap G(k)$, we have that $G(g \wedge k) \subseteq G(g) \cap G(k)$. For the converse, suppose that $h \in G(g) \cap G(k)$, and so $|h| \leq ng$ and $|h| \leq nk$. But then $|h| \leq ng \wedge mk \leq n(g \wedge mk) \leq nm(g \wedge k)$, by 1.1.5, and so $h \in G(g \wedge k)$.

Finally, suppose that $C \in \mathcal{C}(G)$ is compact. Now, $C = \bigvee\{G(g) : g \in C\}$, and so $C = G(g_1) \vee \ldots \vee G(g_n)$, for $\{g_i\} \subseteq C$. Now without loss of generality each g_i is positive; but then $C = G(g_1 \vee \ldots \vee g_n)$. //

It is in general more difficult to describe the ℓ-subgroup generated by a collection of elements:

Proposition 1.2.4 Let S be a subset of G. Then the ℓ-subgroup of G generated by S(i.e., the smallest ℓ-subgroup of G containing S) consists of all elements of the form

$$\bigvee_I \bigwedge_J \sum_K g_{ijk},$$

where I,J and K are finite index sets and for each ijk ,either $g_{ijk} \in S$ or $-g_{ijk} \in S$.

Proof: This follows from repeated application of the distributive laws and DeMorgan's laws. The tedious details are left to the reader. //

Since $\mathcal{C}(G)$ is a Brouwerian lattice, it is *pseudo-complemented* ; that is, for each $C \in \mathcal{C}(G)$, there exists a unique maximum convex ℓ-subgroup C' for which $C \cap C' = \{0\}$ (see [B]). This is obvious, since $C' = \bigvee\{D \in \mathcal{C}(G) : D \cap C = \{0\}\}$. This complementation operator, as for any Brouwerian lattice, defines a Galois connection, satisfying the following laws:

(a) $C \subseteq C''$.
(b) If $B \subseteq C$, then $B' \supseteq C'$.
(c) $C' = C'''$.
(d) $(B \vee C)' = B' \cap C'$.

We call the convex ℓ-subgroups C satisfying the equation $C = C''$ the *polar* subgroups of G and denote the collection of such by $\mathcal{P}(G)$. The following theorem summarizes the properties of $\mathcal{P}(G)$:

Theorem 1.2.5: $\mathcal{P}(G)$ is a complete Boolean algebra, when equipped with the meet $\cap$, a new join operation $\sqcup$ defined by $B \sqcup C = (B \vee C)''$ and complementation $'$. Furthermore, the map $A \mapsto A''$ is a lattice homomorphism of $\mathcal{C}(G)$ onto $\mathcal{P}(G)$.

We omit the proof of the theorem, since it is basically a theorem about lattices (see [B]). These results in the context of ℓ-groups were originally described by Šik [60] but as a theorem about Brouwerian lattices were outlined considerably before by Glivenko [29].

Notice that although $\mathcal{P}(G)$ is a subset of $\mathcal{C}(G)$, it is by no means a sublattice, since the join operations are distinct. We can say this in lattice-theoretic language by remarking that $\mathcal{C}(G)$ is pseudo-complemented, but not necessarily complemented.

In particular, $C \vee C'$ need not equal G, while $C \sqcup C'$ does. If $C \vee C'$ does equal G, then $C = C''$ and we call C and C' *cardinal summands* of G; we denote this by writing $G = C \boxplus C'$. In this case G is a group direct sum of C and C' , and $g \in G$ is positive exactly when its projections on C and C' are. Notice that the complementary summand of a cardinal summand C is uniquely determined, namely as C' ; of course, this contrasts with general group theory, where complements of summands need not be uniquely determined.

It is important to recognize that the polars can also be described in terms of the lattice of G:

Proposition 1.2.6 Suppose X is any subset of an ℓ-group G. Then

$$\{g \in G : |g| \wedge |x| = 0, \text{ for all } x \in X\}$$

is a polar subgroup of G. Furthermore, if X is a convex ℓ-subgroup, then this subgroup is precisely X' .

Proof: Let $Y = \{g \in G : |g| \wedge |x| = 0, \text{ for all } x \in X\}$, and suppose that $g, h \in Y^+$. Then, if $x \in X$,

$$0 = |x| \wedge g = |x| \wedge (|x| \wedge h) + g) = |x| \wedge (|x| + g) \wedge (h + g) = |x| \wedge (h + g),$$

and so $h+g \in Y$. Since Y is obviously closed under additive inverses, and is convex, Y is a convex ℓ-subgroup. Now let $C = \bigvee\{G(x) : x \in X\}$. From Proposition 1.2.5 it is clear that $Y \cap G(x) = \{0\}$ for all $x \in X$ and so $Y \cap C = \{0\}$. Thus $Y \subseteq C'$. If we choose $k \in Y$, then there exists $x \in X$ with $|k| \wedge |x| > 0$. Replacing k by $|k| \wedge |x|$ if necessary, we may suppose that $k \leq |x|$, and so $k \in C(x)$. Thus $x \notin C'$ and hence $Y = C'$. //

Because of this proposition, we shall not hesitate to use the notation X' even if X is not a convex ℓ-subgroup. Also, we shall understand by g' and g'' the polar subgroups $\{g\}'$ and $\{g\}''$. The polar subgroups g'' are called *principal polars.*

We shall now consider a natural generalization of polar subgroups. A convex ℓ-subgroup is *closed* if it is closed with respect to infinite meets and joins which exist in the ℓ-group. More precisely, a convex ℓ-subgroup C of an ℓ-group G is closed with respect to infinite joins if whenever $\{c_\lambda\} \subseteq C$, and $c = \bigvee c_\lambda$ exists in G, then $c \in C$. Note that the infinite version of DeMorgan's laws implies that we really need only require that the convex ℓ-subgroup be closed with respect to joins. In fact, an easy argument using the other infinite distributive laws shows that a convex ℓ-subgroup is closed exactly if its positive cone is closed with respect to infinite joins. We shall denote the set of closed convex ℓ-subgroups of an ℓ-group G by $\mathcal{K}(G)$.

We now observe that every polar is closed: for if P is a polar with $\{c_\lambda\} \subseteq P^+$ and $g \in P'$, then $g \wedge \bigvee c_\lambda = \bigvee(g \wedge c_\lambda) = 0$. Furthermore, every convex subgroup of an o-group is closed, as the reader may easily verify. This shows that not all closed convex ℓ-subgroups need be polars, since an o-group has only the two trivial polars. To see that not all convex ℓ-subgroups need be closed, see E12.

It is obvious that the intersection of a set of closed convex ℓ-subgroups is closed, and since the whole ℓ-group is obviously closed, we then have a unique minimum closed convex ℓ-subgroup containing any given convex ℓ-subgroup C, which we shall call the *closure* of C, and denote by $\overline{C}$. In fact, it is a straightforward matter to prove that the positive cone of the closure is given by $\overline{C}^+ = \{g \in G : g = \bigvee c_\lambda$, where $\{c_\lambda\} \subseteq C^+\}$. We can now define the join of two closed convex ℓ-subgroups C and K by $\overline{C \vee K}$. The reader may verify that this operation, together with ordinary intersection, makes $\mathcal{K}(G)$ a complete Brouwerian lattice, which need not be a sublattice of $\mathcal{C}(G)$ (see E12); furthermore, $\mathcal{P}(G)$ need not be a sublattice of $\mathcal{K}(G)$(see E38).

Notice that the polars and the principal convex ℓ-subgroups are distinguishable in lattice-theoretic terms within the lattice of convex ℓ-subgroups; we shall discover below that this is also true for important classes of subgroups called prime and regular. However, no such characterization of closed convex ℓ-subgroups is possible, as E53 reveals.

We can characterize closed convex ℓ-subgroups nicely in terms of homomorphisms. We call a lattice homomorphism *complete* if it preserves all (not necessarily finite) meets and joins. The following theorem is due to Byrd [67] and Weinberg [62]:

Theorem 1.2.7 A convex ℓ-subgroup C of an ℓ-group G is closed if and only if the natural lattice homomorphism from G onto its lattice G/C of right cosets is complete.

Proof: First suppose that the natural lattice homomorphism is complete; let $\{c_\lambda\} \subseteq C$, and suppose that $c = \bigvee c_\lambda$ exists in G. Then $C + c = C + \bigvee c_\lambda = \bigvee(C + c_\lambda) = C + 0$, and so $c \in C$, as required.

Conversely, suppose that C is closed. Clearly, we need only prove that the natural homomorphism preserves infinite joins. For this purpose, let $\{g_\lambda\} \subseteq G$, and suppose that $\bigvee g_\lambda$ exists in G. We must show that this join is preserved by the natural lattice homomorphism onto G/C. Clearly $C + g \geq C + g_\lambda$, for all λ. Suppose by way of contradiction $C + g > C + h \geq C + g_\lambda$, for all λ. Then $C + g \vee h = (C + g) \vee (C + h) = C + g > C + h$, and so $(g - h) \vee 0 \notin C$. We have that $C + g_\lambda \vee h = C + h$, and so $(g_\lambda - h) \vee 0 \in C$, for all λ. But then $(g - h) \vee 0 = \bigvee\big((g_\lambda - h) \vee 0\big) \in C$, since C is closed. This contradiction proves the theorem. //

We now need another definition from lattice theory: An element x of a lattice is *meet-irreducible* if whenever $x = \bigwedge_\Lambda x_\lambda$, then $x = x_\lambda$, for some λ. (An element x is *finitely meet-irreducible* if x satisfies the former when Λ is finite.) We shall now examine the meet-irreducible elements of the lattice $\mathcal{C}(G)$, and in particular show that any element of this lattice can be obtained as the intersection of such in the following theorem due to Conrad [65].

Theorem 1.2.8 Let $P \in \mathcal{C}(G)$. Then P is meet-irreducible in $\mathcal{C}(G)$ if and only if P is maximal in $\mathcal{C}(G)$ with respect to not containing some $g \in G$.

Proof: Suppose P is meet-irreducible. Let $P^* = \bigcap\{C \in \mathcal{C}(G) : C \supset P\}$. Since P is meet-irreducible, $P^* \supset P$. Choose any $g \in P^* \backslash P$. Then, $P \subset P \vee G(g) \subset P^*$, and so $P \vee G(g) = P^*$. Thus P is maximal with respect to not containing g.

Conversely, if P is maximal with respect to not containing g, then $g \in C$ if $C \supset P$. But then $\bigcap\{C \in \mathcal{C}(G) : C \supset P\} \supset P$. //

We call a convex ℓ-subgroup with the properties of Theorem 1.2.8 *regular*, and denote the set of them by $\Gamma(G)$. If $P \in \Gamma(G)$, then P^* is the *cover* of P, and if $g \in P^* \backslash P$, then P is a *value* of g. (This terminology arose from the theory of valuations of fields in totally ordered groups (see [Ja], [Gi]).)

Notice that if P is a value of g, then the conjugate $-h+P+h$ is a value of $-h+g+h$; that is, the set $\Gamma(G)$ is closed under conjugation. Zorn's lemma gives us the following:

Proposition 1.2.9 Every element of an ℓ-group has at least one value. Consequently, each convex ℓ-subgroup can be obtained as the intersection of regular subgroups.

A regular subgroup possesses the further property that its set of right cosets is totally ordered; convex ℓ-subgroups with this property are of profound importance in the theory of ℓ-groups. They are called *prime* subgroups, and are characterized by the following theorem. Conrad [65] proved this theorem for arbitrary ℓ-groups, thus generalizing earlier work of Conrad, Harvey and Holland [63] for abelian ℓ-groups and Johnson and Kist [62] for vector lattices.

Theorem 1.2.10 For $P \in \mathcal{C}(G)$, the following are equivalent:

(a) P is prime (that is, G/P is totally ordered).
(b) $\{C \in \mathcal{C}(G) : C \supseteq P\}$ is totally ordered under inclusion.
(c) P is finitely meet-irreducible in $\mathcal{C}(G)$.
(d) If $a \wedge b \in P$ then $a \in P$ or $b \in P$.
(e) If $a \wedge b = 0$ then $a \in P$ or $b \in P$.

Consequently, each regular subgroup is prime.

Proof: (a) $\Rightarrow$ (b). Suppose that B and C are incomparable convex ℓ-subgroups containing P. Choose $b \in B^+ \backslash C$ and $c \in C^+ \backslash B$. Without loss of generality $P+b \geq P+c$. Then there exists $p \in P$ with $p+b \geq c$. Hence $c \in B$, a contradiction.

(b) $\Rightarrow$ (c). If $B \cap C = P$, then $B, C \supseteq P$ and so without loss of generality $B \supseteq C$. Thus $P = C$.

(c) $\Rightarrow$ (d). Now, $(P \vee G(a)) \cap (P \vee G(b)) = P \vee (G(a) \cap G(b)) = P \vee (G(a \wedge b)) = P$, and so either $P = P \vee G(a)$ or $P = P \vee G(b)$, and thus $a \in P$ or $b \in P$.

(d) $\Rightarrow$ (e) is clear.

(e) $\Rightarrow$ (a). Given $g, h \in G$, we have

$$((g \vee h) - g) \wedge ((g \vee h) - h) = (g \vee h) + (-g \wedge -h) = (g \vee h) + -(g \vee h) = 0,$$

and so (without loss of generality) $g \wedge h - h \in P$. Then, $P + h = P + (g \vee h) \geq P + g$.

That each regular subgroup is prime follows immediately from (c). //

Notice that it follows that if G is an o-group then $\{0\}$ is prime and so the convex ℓ-subgroups of G form a chain.

It is the similarity of condition (d) to the ring-theoretic definition which undoubtedly motivated the use of the term "prime". Since all representation theorems for ℓ-groups describe them in terms of totally ordered structures, it is not surprising that the primes will play a key role. Notice that (b) implies that the set of primes of an ℓ-group (and a fortiori, $\Gamma(G)$ as well) forms a root system (recall the definition from 1.1.12).

From (b) it is apparent that the intersection of a chain of primes is still prime. Since each prime may evidently be placed in a maximal chain of primes, it follows that each prime contains a (not necessarily unique) minimal prime. With minimal primes comes an important connection between the notions of prime and polar, as the next two theorems will reveal.

It is apparent from (d) of Theorem 1.2.10 that if P is a prime, then the set $F = G^+ \backslash P$ satisfies the property that $a \wedge b > 0$ for all $a, b \in F$. A subset F of G^+ maximal with respect to this property is called an *ultrafilter*. We can now characterize minimal primes in the following theorem due to Johnson and Kist [62] , Banaschewski [64] and in its most general form to Conrad and McAlister [69].

Theorem 1.2.11 For $P \in \mathcal{C}(G)$, the following are equivalent:

(a) P is a minimal prime.

(b) $G^+ \backslash P$ is an ultrafilter.

(c) $P = \bigcup\{g' : g \notin P\}$.

(d) P is a prime, and for all $h \in P, h' \not\subseteq P$.

Proof: (a) $\Rightarrow$ (b). We have already observed that for all $a, b \in G^+ \backslash P, a \wedge b > 0$. By Zorn's lemma we may obtain an ultrafilter F which contains $G^+ \backslash P$. For each $g \in F$, it is evident that $(g')^+ \subseteq G^+ \backslash F \subseteq P$. But then $Q = \bigcup\{g' : g \in F\} \subseteq P$. We will now show that Q is a subgroup. If $a \in g'$, $b \in h'$ with $g, h \in F$, then

$$g \wedge h \wedge |a + b| \leq g \wedge h \wedge (|a| + |b| + |a|) \leq (g \wedge h \wedge |a|) + (g \wedge h \wedge |b|) + (g \wedge h \wedge |a|) = 0,$$

and so $a + b \in (g \wedge h)'$. Since $g \wedge h \in F$ we have $a + b \in Q$. Thus Q is clearly a convex ℓ-subgroup; to show that it is prime, suppose that $c \wedge d = 0$ and $c \notin Q$. Then $c \wedge g > 0$ for all $g \in F$. But since F is an ultrafilter, $c \in F$ and so $d \in c' \subseteq Q$. Since P is a minimal prime, $P = Q$ and $F = G^+ \backslash P$.

(b) ⇒ (c). The proof above shows that if $G^+\backslash P$ is an ultrafilter , $\bigcup\{g' : g \in P\}$ is a prime subgroup which is contained in P. If $0 < h \notin \bigcup\{g' : g \in P\}$ then $h \wedge g > 0$, for all $g \in G^+\backslash P$; since $G^+\backslash P$ is an ultrafilter, $h \in P$.

(c) ⇒ (d). This is obvious.

(d) ⇒ (a). Suppose $Q \in \mathcal{C}(G)$ and $Q \subset P$. Choose $h \in P^+\backslash Q$. Then there exists $g \in P$ with $h \wedge g = 0$. But then Q can't be prime. //

As a corollary we obtain the following:

Corollary 1.2.12 Each polar is an intersection of minimal primes, specifically the intersection of all those minimal primes not containing the complement of the polar.

Proof: Let $A \in \mathcal{P}(G)$, and suppose P is a minimal prime not containing A' . Then there exists $a \in (A')^+\backslash P$, and $A \subseteq a' \subseteq P$. Thus the intersection of such minimal primes contains A. Now, if $0 < a \notin A$, there exists $b \in A'$ with $a \wedge b > 0$ and so we may choose an ultrafilter F containing a and b. Then the complementary minimal prime P does not contain a. Since a was arbitrary, the intersection of all such minimal primes must equal A. //

Finally, we obtain a theorem which relates the primes of an ℓ-group to the primes of a convex ℓ-subgroup:

Theorem 1.2.13 Let G be an ℓ-group and H a convex ℓ-subgroup . Then the mapping $\Phi : P \mapsto H \cap P$ gives a one-to-one correspondence between the prime subgroups of G not containing H and the proper prime subgroups of H.

Proof: It is obvious that if P is a prime subgroup of G not containing H, then $P \cap H$ is a proper prime subgroup of H. To show that this establishes a one-to-one correspondence, we shall define the inverse map of Φ:

For Q a prime subgroup of H, let $Q\Psi = \bigvee\{R \in \mathcal{C}(G) : R \cap H = Q\}$. We must show that $Q\Psi$ is a prime subgroup of G. Toward that end suppose that $a, b \in G$ and $a \wedge b = 0$. Consider $(G(a) \vee Q) \cap H$ and $(G(b) \vee Q) \cap H$. Since these are both convex ℓ-subgroups of H which contain the prime Q, we have (without loss of generality) $(G(a) \vee Q) \cap H \supseteq (G(b) \vee Q) \cap H \supseteq Q$. But

$$(G(a) \vee Q) \cap H \cap (G(b) \vee Q) \cap H = H \cap \Big(Q \vee (G(a) \cap G(b))\Big) = H \cap Q = Q.$$

Thus $(G(b) \vee Q) \cap H = H$, and so $b \in Q\Psi$.

It remains to show that Φ and Ψ are inverse maps. But

$$Q\Psi\Phi = H \cap \bigvee\{R : R \cap H = Q\} = \bigvee\{H \cap R : R \cap H = Q\} = Q.$$

On the other hand, it is obvious that $P\Phi\Psi \supseteq P$. Conversely if $g \in (P\Phi\Psi)^+$, choose $h \in H^+\backslash P$. Then $g \wedge h \in P\Phi\Psi \cap H = P \cap H$. But P is prime and so $g \in P$ as required. //

Chapter 2: *Bernau's Representation for Archimedean ℓ-groups*

In the next four chapters we shall prove the four fundamental representation theorems for lattice-ordered groups, in increasing order of generality. Consequently, we shall begin with the representation of archimedean ℓ-groups, as continuous almost finite real-valued functions on a topological space. It was Bernau [65b] who first stated and proved this theorem in the ℓ-group context, but because of its analytic flavor it should not be surprising that this result was anticipated in work on partially ordered vector spaces, done by Amemiya [53], Nakano [N], Vulich [V] and Pinsker [49], and others.

An ℓ-group is *archimedean* if for any elements g, h, $ng \leq h$ for all positive integers n implies that $g \leq 0$. Note that we may as well suppose that g and h are positive in this definition. The first important observation to be made regarding such ℓ-groups is that they are necessarily abelian (Theorem 2.2). The proof of this theorem given here uses only elementary facts about ℓ-groups and is due to Bernau [65b]; we give an alternate and easier to follow proof of this fact at the end of section 4.2 but that proof requires some theorems concerning representable and normal-valued ℓ-groups given in chapter 4. The first-time reader is advised to postpone the proof of Theorem 2.2 until then. Birkhoff's original proof [B] depended on embedding G in a cut completion (see Chapter 8), and then citing Iwasawa's result [43] that complete groups are abelian.

Lemma 2.1 If G is archimedean and $g \in G$ then g' is normal.

Proof: Suppose that $g \wedge h = 0$ and $k \in G^+$. We must show that if $a = g \wedge (-k + h + k)$, then $a = 0$. Let n be any positive integer. Since $g \wedge h = 0$, we have that

$$a \wedge (k + a - k) = g \wedge (-k + h + k) \wedge (k + g - k) \wedge h = 0$$

and so

$$0 = na \wedge n(k + a - k) = na \wedge (k + na - k) \geq (na - k) \vee 0,$$

where the last inequality holds because each term in the join is bounded above by each term in the meet. Thus, $(na - k) \vee 0 = 0$, and so $na \leq k$ for all positive integers n and since G is archimedean, $a = 0$. A similar argument shows that $g \wedge (k + h - k) = 0$, and so g' is normal. //

Theorem 2.2 Each archimedean ℓ-group is abelian.

Proof: Suppose that G is an archimedean ℓ-group, and $g \in G^+$. By the above lemma, g' is normal. Since $g'' = \bigcap\{b' : b \in g'\}$, it is clear that g'' is normal too.

We next claim that the ℓ-group G/g'' is still archimedean. For suppose that $g'' + x$, $g'' + y \in (G/g'')^+$, and $n(g'' + x) \leq g'' + y$, for all positive n. Then there exist $a_n \in (g'')^+$, with $nx \leq a_n + y$. Choose any $h \in (g')^+$. Then

$$n(x \wedge h) \leq nx \wedge nh \leq (a_n + y) \wedge nh \leq (a_n \wedge h) + (y \wedge nh) = 0 + (y \wedge nh) \leq y,$$

and so $x \wedge h = 0$. Thus, $x \in g''$ and so $g'' + x = g'' + 0$.

Now, to show that G is abelian, we evidently need only prove that positive elements commute. Choose two such elements a, b. Let $t = a+b+b+a-b-a-a-b$. Now $t = t^+ - t^-$ and $t^- \in (t^-)''$, and so $t \geq 0$ modulo $(t^-)''$. It follows that, modulo $(t^-)''$,

(1) $$a+b-a-b \leq -a-b+a+b$$

and

(2) $$-b-a+b+a \leq b+a-b-a.$$

We now use these two inequalities to show inductively that

(3) $$n(-a-b+a+b) \leq n(-a-b)+n(a+b)$$

and

(4) $$n(b+a-b-a) \leq n(b+a)+n(-b-a),$$

for all positive integers n, modulo $(t^-)''$. Both inequalities are obvious for $n=1$. If we assume the inequality (3) for $n-1$, we obtain

$$\begin{aligned} n(-a-b+a+b) &= -a-b+a+b+(n-1)(-a-b+a+b) \\ &\leq -a-b+a+b+(n-1)(-a-b)+(n-1)(a+b) \\ &\leq 2(-a-b)+(a+b)+(n-2)(-a-b)+(n-1)(a+b) \\ &\leq \cdots \leq n(-a-b)+n(a+b), \end{aligned}$$

by applying inequalities (1) and (2) repeatedly. The other inequality follows similarly.

For positive integer n, since

$$(n+1)(b+a) = b+n(a+b)+a \geq n(a+b)$$

and

$$(n+1)(a+b) = a+n(b+a)+b \geq n(b+a),$$

it follows that

(5) $$n(-a-b)+n(a+b) \leq b+a$$

and

(6) $$n(b+a)+n(-b-a) \leq a+b.$$

Thus, combining inequalities (1), (2), (5) and (6), we have (modulo $(t^-)''$)

$$n(-a-b+a+b) \leq b+a \text{ and } n(b+a-b-a) \leq a+b.$$

Since $G/(t^-)''$ is archimedean, it follows from the first inequality that modulo $(t^-)''$, $-a-b+a+b \leq 0$, that is, $a+b \leq b+a$. Similarly, the second inequality yields $b+a \leq a+b$ and so $a+b=b+a$, modulo $(t^-)''$.

The same kind of argument can be constructed to show that $a+b=b+a$, modulo $(t^+)''$. Since $(t^+)'' \cap (t^-)'' = \{0\}$, this means that $a+b=b+a$; that is, G is abelian. //

Note, however, that not every abelian ℓ-group is archimedean as $\mathbf{Z} \oplus \mathbf{Z}$ ordered lexicographically easily shows.

The first step toward a representation theorem for archimedean ℓ-groups was taken by Hölder [01], who took care of the totally ordered case by proving the following theorem. His proof did of course include a proof of the totally ordered version of the previous theorem.

Theorem 2.3 Let G be a totally ordered group. Then the following are equivalent:

(a) G is archimedean.

(b) G is o-isomorphic to a subgroup of the additive reals.

(c) G has no proper convex subgroups.

Proof: (a) ⇒ (b). Choose $g \in G^+$. Then define $\phi : G \to \mathbf{R}$ by

$$\phi(h) = \bigwedge\left\{\frac{m}{n} : mg > nh, \text{ with } m, n \in \mathbf{Z}\right\}.$$

Because G is archimedean and totally ordered, there exist positive integers p, q for which $pg > |h|$ and $q|h| > g$, and so the set of rational numbers in the infimum above is both nonempty and bounded below; thus ϕ is well-defined.

To show that ϕ is a homomorphism, suppose that for h, $k \in G$ and $m, n, p, q \in \mathbf{Z}$, $mg > nh$ and $pg > qk$. Then $(mq+np)g > qnh + nqk = qn(h+k)$, since G is abelian. Therefore,

$$\frac{m}{n} + \frac{p}{q} = \frac{mq+np}{qn} \geq \phi(h+k),$$

and since this is true for all such m,n,p and q we have $\phi(h+k) \leq \phi(h) + \phi(k)$. In a similar way, if $\frac{m}{n} \leq \phi(h)$ and $\frac{p}{q} \leq \phi(k)$, it is easy to see that $\frac{m}{n} + \frac{p}{q} \leq \phi(h+k)$, and so $\phi(h+k) = \phi(h) + \phi(k)$. It is evident that ϕ preserves order, and so it remains to show that ϕ is one-to-one. If $b > 0$, then there exists positive integer n for which $nb \geq g$, and so $\phi(b) \geq \frac{1}{n}$; similarly, if $b < 0$, $\phi(b) \leq \frac{-1}{n}$, for some n, and so ϕ is one-to-one.

(b) ⇒ (c). This is obvious.

(c) ⇒ (a). This follows immediately from Proposition 1.2.3. //

We shall begin our discussion of the representation theorem for an arbitrary archimedean ℓ-group by examining the topological space that will be used. As we have seen, for any ℓ-group G the set $\mathcal{P}(G)$ of polars of G is a complete Boolean algebra. Let $X(G)$ (or just X) denote the topological space corresponding to $\mathcal{P}(G)$ according to Stone's duality theorem [40].

More specifically, a *filter* on $\mathcal{P}(G)$ is a subset of $\mathcal{P}(G) \backslash \{\{0\}\}$ closed under finite intersection and supersets. Then we let $X(G)$ be the set of all ultrafilters (i.e., maximal filters) on $\mathcal{P}(G)$. Given any $P \in \mathcal{P}(G)$, let $\mathcal{O}(P) = \{x \in X : P \in x\}$. The set of all such $\mathcal{O}(P)$ forms a subbase for the topology on X . In this case X is a compact space, and the closure of each open set is still open; this latter property makes X *extremally disconnected.* An open set is said to be *regularly open* if it is the interior of its closure, and the set of all regularly open sets of any topological space is a compete Boolean algebra [GJ]. For the X constructed above each regularly open set is *clopen* (both open and closed), and this complete Boolean algebra is isomorphic to $\mathcal{P}(G)$ under the mapping $P \mapsto \mathcal{O}(P)$. For our purposes, a crucial property of an extremally disconnected space is that every open subset and every dense subset is *C^*-embedded*; that is, each bounded continuous real-valued function on such a subset admits a (necessarily unique) continuous extension to all of X [GJ].

The target for our representation theorem will then be an ℓ-group of the following form: Let X be an extremally disconnected compact space, and let $D(X)$ be the set of all continuous functions f from X into $\mathbf{R} \cup \{-\infty, +\infty\}$ (the two point compactification of the real numbers), for which $f^{-1}(\mathbf{R})$ is dense (and necessarily open). We shall describe such functions as "continuous almost finite real-valued".

We first observe that $D(X)$ is in fact an additive group. For, if $g, h \in D(X)$, then $g^{-1}(\mathbf{R})$ and $h^{-1}(\mathbf{R})$ are open dense sets and consequently so is their intersection. But then $g+h$ is defined on $g^{-1}(\mathbf{R}) \cap h^{-1}(\mathbf{R})$, and since this is an open dense set, this sum admits a unique extension to all of X ; we shall call this extension $g+h$. It is easily verified that this addition, together with the pointwise order (where $-\infty < r < +\infty$, for all $r \in \mathbf{R}$) makes $D(X)$ an archimedean ℓ-group. Furthermore, since $g \wedge h = 0$ exactly when their supports are disjoint, the Boolean algebra of polars for $D(X)$ is isomorphic to the Boolean algebra of clopen subsets of X .

We are now ready to state and prove Bernau's representation theorem for archimedean ℓ-groups:

Theorem 2.4 Let G be an archimedean ℓ-group. Then G may be ℓ-embedded into $D(X)$, where X is the Stone space corresponding to the Boolean algebra $\mathcal{P}(G)$.

Proof: First, we choose by Zorn's lemma a maximal pairwise disjoint set $\{g_\alpha\}$ of elements from G^+ . For any $f \in G^+$, we are now ready to define a function $\overline{f}$ on X as follows. Let $x \in X$; then

$$\overline{f}(x) = \bigwedge \Big\{ \frac{m}{n} : (mg_\alpha - nf)^{+\prime\prime} \in x, \text{ for some } \alpha \Big\}.$$

(We understand that this infimum is $+\infty$, if the set defined in the definition is empty).

We claim that for any positive integer n ,

$$\bigsqcup_{m>0} (mg_\alpha - nf)^{+\prime\prime} = g_\alpha''.$$

For if not, there exists $\{0\} \neq h'' \subseteq g_\alpha'' \setminus \bigsqcup (mg_\alpha - nf)^{+\prime\prime}$. Then $h \in (mg_\alpha - nf)^{+\prime}$ and so, $m(g_\alpha \wedge h) \leq nf$, for all m . But G is archimedean, and so $g_\alpha \wedge h = 0$, which is a contradiction.

We can now verify that $\overline{f} \in D(X)$. If $\overline{f}^{-1}(\mathbf{R})$ is not dense, there exists an open set upon which $\overline{f}$ is $+\infty$; we may assume that this set is of the form $\mathcal{O}(h'')$, for $h \in G^+$. It follows that $(mg_\alpha - nf)^{+''} \cap h'' = \{0\}$, for all positive integers m and n and any α . But then $h'' \cap g''_\alpha = \{0\}$ for all α , which contradicts the maximal disjointness of the g_α's .

We must now check that $\overline{f}$ is continuous. If r is a positive real number, then

$$\overline{f}^{-1}(r,\infty) = \bigcup\left\{\mathcal{O}((mg_\alpha - nf)^{+''}) : r < \frac{m}{n}\right\},$$

which is clearly open, while

$$\overline{f}^{-1}[r,\infty) = \bigcap\left\{\mathcal{O}((mg_\alpha - nf)^{+''}) : r > \frac{m}{n}\right\},$$

which is clearly closed.

We shall now prove that $\overline{f}+\overline{h} = \overline{f+h}$. We need only show that $\overline{f}(x)+\overline{h}(x) = \overline{f+h}(x)$, for all $x \in \overline{f}^{-1}(\mathbf{R}) \cap \overline{h}^{-1}(\mathbf{R})$. Suppose first of all that $(m_1g_\alpha - n_1f)^{+''}$ and $(m_2g_\alpha - n_2f)^{+''}$ are elements of x . Since $g''_\alpha \cap g''_\beta = \{0\}$ if $\alpha \neq \beta$, it is clear that $\alpha = \beta$. Because $(na)'' = a''$, we may assume that $n_1 = n_2$. Thus we have, by Proposition 1.1.8, that

$$\begin{aligned} 0 &< (m_1g_\alpha - nf)^+ \wedge (m_2g_\alpha - nh)^+ \\ &\leq (m_1g_\alpha - nf + m_2g_\alpha - nh)^+ \\ &= ((m_1 + m_2)g_\alpha - n(f+h))^+. \end{aligned}$$

Consequently,

$$(m_1g_\alpha - nf)^{+''} \cap (m_2g_\alpha - nh)^{+''} \subseteq ((m_1+m_2)g_\alpha - n(f+h))^{+''},$$

and so $\overline{f}(x) + \overline{h}(x) \geq \overline{f+h}(x)$.

For the reverse inequality, suppose that $\frac{m_1}{n_1} < \overline{f}(x)$ and $\frac{m_2}{n_2} < \overline{h}(x)$. This means that $(m_1g_\alpha - n_1f)^{+''}$ and $(m_2g_\alpha - n_2h)^{+''}$ are not elements of x , for all α . Consequently, for all α, the polar

$$((m_1n_2g_\alpha - n_1n_2f)^+ \vee (n_1m_2g_\alpha - n_1n_2h)^+)'' = ((m_1n_2g_\alpha - n_1n_2f)^+ + (n_1m_2g_\alpha - n_1n_2h)^+)''$$

is not an element of x . But

$$\begin{aligned} &(m_1n_2g_\alpha - n_1n_2f)^+ + (n_1m_2g_\alpha - n_1n_2h)^+ \\ &\geq (m_1n_2g_\alpha - n_1n_2(f+h))^+ + (n_1m_2g_\alpha - n_1n_2(f+h))^+ \\ &\geq ((m_1n_2 + n_1m_2)g_\alpha - n_1n_2(f+h))^+, \end{aligned}$$

and so $((m_1n_2 + n_1m_2)g_\alpha - n_1n_2(f+h))^{+''}$ is not an element of x . That is, $\frac{m_1}{n_1} + \frac{m_2}{n_2} \leq \overline{f+h}(x)$. Thus, $\overline{f}(x) + \overline{h}(x) = \overline{f+h}(x)$.

It is a straightforward matter to check that the map we have defined is one-to-one, and a lattice homomorphism, from G^+ into $D(X)^+$. This map then admits a unique extension to an ℓ-monomorphism from G into $D(X)$. //

The representation theorem which we have just proved bears several advantages which we shall further explore in later chapters. What is crucial is that the representation space is uniquely determined by the Boolean algebra of polars. The meaning of this, as we shall discover in Chapter 8, is that the target ℓ-group $D(X)$ can be described in purely algebraic terms from G .

The only apparent drawback of this representation is that the functions used are not everywhere finite. However, not all archimedean ℓ-groups can be represented as subdirect products of reals, as E21 shows. Any archimedean ℓ-group can be represented as a group of real-valued functions (on a partially ordered set), but only at the expense of leaving the class of archimedean ℓ-groups. This sort of representation theorem, applicable to any abelian ℓ-group, is the subject of the next chapter.

Chapter 3: *The Conrad-Harvey-Holland Representation*

In 1907 H. Hahn [07] proved that every abelian totally ordered group could be represented as a group of real-valued functions on a totally ordered set. His proof is long and difficult, and is one of the first successful uses of the technique of transfinite induction. It was not until after the Second World War that simplified proofs of Hahn's theorem appeared, by Conrad [53] and Clifford [54]. In 1963, Conrad, Harvey and Holland [63] were able to generalize Hahn's theorem to the class of lattice-ordered groups, using the idea of Banaschewski's proof [57] of the original theorem; their theorem states that any abelian ℓ-group can be represented as a group of real-valued functions on a partially ordered set (in fact, on a root system). In this chapter we shall present Wolfenstein's proof [66] (or see [C]) of the Conrad-Harvey-Holland theorem.

Given an abelian ℓ-group G, the set $\Gamma(G)$ of regular subgroups of G is a root system (see 1.1.13 and the remarks following Theorem 1.2.10). Consequently, it makes sense to speak of the abelian ℓ-group $V(\Gamma(G), \mathbf{R})$; this is the target for our representation theorem. In this chapter we shall write elements of $\Gamma(G)$ as γ or G_γ, depending on whether we wish to emphasize $\Gamma(G)$'s role as an index set or as the set of regular subgroups of G. As usual, we shall denote the cover of $G_\gamma \in \Gamma(G)$ by G_γ^*.

For any $G_\gamma \in \Gamma(G)$, G_γ^*/G_γ is by Hölder's theorem (2.2) isomorphic to a subgroup of the real numbers $\mathbf{R}$; choose such an isomorphism. We shall then define an ℓ-homomorphism π_γ by the composition

$$G_\gamma^* \longrightarrow G_\gamma^*/G_\gamma \longrightarrow \mathbf{R}$$

where the first map is the natural map, and the second is the monomorphism we have chosen.

The remainder of the proof consists in stitching together these component maps in such a way that the order and group structures are preserved. We shall accomplish this by inductively extending such maps to larger and larger subgroups. We need the following terminology to accomplish this: Given a subgroup H of G, a *π-mapping* is a group homomorphism $\alpha : H \to V(\Gamma(G), \mathbf{R})$ such that $g \in G_\gamma^* \cap H$ implies that $(g\alpha)_\gamma = g\pi_\gamma$.

Note that a π-mapping is not necessarily an ℓ-homomorphism. However, the following lemma guarantees that a π-mapping defined on all of G is in fact an ℓ-isomorphism:

Lemma 3.1 Let α be a π-mapping of G into $V(\Gamma(G), \mathbf{R})$ and $g \in G$. Then G_γ is a value for g if and only if γ is a maximal element of $spt(g\alpha)$. Consequently α is an ℓ-monomorphism.

Proof: Let G_γ be a value of g; clearly $(g\alpha)_\gamma = g\pi_\gamma \neq 0$. If $\delta > \gamma$ then $g \in G_\delta$, and so $(g\alpha)_\delta = g\pi_\delta = 0$. Thus, γ is a maximal element of $spt(g\alpha)$.

Conversely, suppose that γ is a maximal element of $spt(g\alpha)$ and so $g \notin G_\gamma$. If $g \notin G_\gamma^*$ then δ is a value for g, for some $\delta > \gamma$ and so $\delta \in spt(g\alpha)$, which is a contradiction. Thus, $g \in G_\gamma^*/G_\gamma$.

We must now show that α is an ℓ-monomorphism. Now, α is clearly one-to-one and is also order-preserving, since each π_γ is. We shall now show that α preserves disjointness. For that purpose, suppose that $a \wedge b = 0$. But then no value of a is comparable with a value of b, and so no maximal component of $a\alpha$ is comparable with a maximal component of $b\alpha$. But all of these maximal components are positive, and so $a\alpha \wedge b\alpha = 0$. We can now show that α is an ℓ-homomorphism. For if $a \in G$, then

$$0 \leq (a \vee 0)\alpha - (a\alpha \vee 0) = (0 \vee -a)\alpha \wedge (a \vee 0)\alpha = 0,$$

since α preserves disjointness, and so $(a \vee 0)\alpha = a\alpha \vee 0$. //

Theorem 3.2 If G is an abelian ℓ-group, then G is ℓ-isomorphic to an ℓ-subgroup of $V(\Gamma(G), \mathbf{R})$.

Proof: We shall show that if there exists a π- mapping from a proper subgroup of G into $V(\Gamma(G), \mathbf{R})$ then we can extend this to a π-mapping from a larger subgroup of G into $V(\Gamma(G), \mathbf{R})$. If the subgroup is $\{0\}$, we clearly have such a π-mapping. So we may suppose that there exist a subgroup H of G and a π-mapping α from H into $V(\Gamma(G), \mathbf{R})$. If $H = G$, then the proof is complete by Lemma 3.1. If not, choose $g \in H \backslash G$. We shall first define α on g:

Let $\langle g \rangle$ denote the cyclic subgroup generated by g. If $\langle g \rangle \cap (H + G_\gamma^*) = \{0\}$, then let $(g\alpha)_\gamma = 0$. If this intersection is nonzero, there exists a positive integer n, $h \in H$ and $c \in G_\gamma^*$ with $ng = h + c$; we let $(g\alpha)_\gamma = (\frac{1}{n})((h\alpha)_\gamma + c\pi_\gamma)$.

We first must verify that $g\alpha$ is well-defined. Suppose then that $mg = h_1 + c_1$, where $m \in \mathbf{Z}^+$, $h_1 \in H$ and $c_1 \in G_\gamma^*$. Then $mh - nh_1 = nc_1 - mc \in H \cap G_\gamma^*$. This implies that

$$\begin{aligned} m(h\alpha)_\gamma - n(h_1\alpha) &= ((mh - nh_1)\alpha)_\gamma = (mh - nh_1)\pi_\gamma \\ &= (nc_1 - mc)\pi_\gamma = nc_1\pi_\gamma - mc\pi_\gamma. \end{aligned}$$

Hence

$$m(h\alpha) + m(c\pi_\gamma) = n(h_1\alpha)_\gamma + n(c_1\pi_\gamma), \text{ and so } (\frac{1}{n})((h\alpha)_\gamma + c\pi_\gamma) = (\frac{1}{m})((h_1\alpha)_\gamma + c_1\pi_\gamma).$$

We next must check that $g\alpha \in V(\Gamma(G), \mathbf{R})$, that is, that $spt(g\alpha)$ satisfies the ascending chain condition. If $\gamma \in spt(g\alpha)$ and $\gamma < \delta$ then $ng = h + c \in H + G_\gamma^* \subseteq H + G_\delta$ and so $c\pi_\delta = 0$. Thus, $(g\alpha)_\delta = (\frac{1}{n})(h\alpha)_\delta$. But $spt(h\alpha)$ satisfies the ascending chain condition, and therefore so does $spt(g\alpha)$.

We shall now extend the definition of α to the subgroup $K = H + \langle g \rangle$. So for $h + mg \in H + \langle g \rangle$ define $(h + mg)\alpha = h\alpha + m(g\alpha)$. Since K is not necessarily a direct sum, we must check that this definition is well-defined. First note that if $mg \in H$ for some integer m, then $(g\alpha)_\gamma = (\frac{1}{m})((mg)\alpha)_\gamma$ and so $m(g\alpha) = (mg)\alpha$. Next, if $h + mg = h_1 + ng \in H + \langle g \rangle$, then $h - h_1 \in H$, and so $(h - h_1)\alpha = ((n - m)g)\alpha = (n - m)(g\alpha)$. Therefore, $h\alpha + m(g\alpha) = h_1\alpha + n(g\alpha)$.

Finally, we must check that α is a π-mapping of K into $V(\Gamma(G), \mathbf{R})$. It is clearly a homomorphism, and so we need only prove that if $k = h + ng \in G^*_\gamma \cap K$ then $(k\alpha)_\gamma = k\pi_\gamma$. We consider the two cases depending on whether $\langle g \rangle \cap (H + G^*_\gamma)$ is zero. If it is, then $ng = -h + k \in \langle g \rangle \cap (H + G^*_\gamma)$, and so $n = 0$. Thus, $k = h \in H$; therefore $(k\alpha)\gamma = k\pi_\gamma$. If $\langle g \rangle \cap (H + G^*_\gamma) \neq \{0\}$, then $png = f + c \in H + G^*_\gamma$, for some positive integer p. But then $pk = ph + f + c$, and so $pk - c = ph + f \in G^*_\gamma \cap H$. So,

$$\begin{aligned} p(k\alpha)_\gamma &= (pk\alpha)_\gamma = ((ph + f)\alpha)_\gamma + c\pi_\gamma = (ph + f)\pi_\gamma + c\pi_\gamma \\ &= (ph + f + c)\pi_\gamma = (pk)\pi_\gamma = p(g\pi_\gamma). \end{aligned}$$

Therefore $(g\alpha)_\gamma = g\pi_\gamma$, and so α is a π-mapping.

By continuing to extend α we eventually obtain a π-mapping defined on all of G, which by the lemma gives us the required ℓ-monomorphism. //

Note that as a corollary of the theorem we obtain that every abelian ℓ-group can be ℓ-embedded into a divisible ℓ-group, since ℓ-groups of the form $V(\Gamma, \mathbf{R})$ are divisible. Actually this can be proved directly: for an abelian ℓ-group G, let G^d be its group-theoretic divisible hull. Call an element h of G^d positive if $nh \in G^+$ for some positive integer n; this makes G^d an ℓ-group.

Part of Hahn's achievement in his 1907 paper was to show that the target groups in his representation theorem for abelian o-groups are in fact unique maximal extensions preserving the chain of convex subgroups (when the extensions are restricted to abelian o-groups). The renewed interest in Hahn's theorem in the 1950s which culminated in the Conrad-Harvey-Holland representation theorem brought with it the question of whether this part of Hahn's work could be extended to the (possibly not abelian) ℓ-group case.

The appropriate question generalizing Hahn's work for o-groups would inquire about extensions of ℓ-groups which preserve this lattice. The first result along this line was Hölder's characterization [01] of additive subgroups of the reals as o-groups whose lattice of subgroups consists just of the trivial subgroups (Theorem 2.2). In this case the group of real numbers then serves as a maximal extension preserving this lattice; that is, it is an archimedean closure in the sense more precisely described below.

Given two positive elements g, h of an ℓ-group G, we say that they are *archimedean equivalent* (henceforth *a-equivalent*) if there exist positive integers m, n so that $g \leq mh$ and $h \leq ng$ (we could clearly assume that $m = n$ to simplify the definition). This obviously amounts to saying that the principal convex ℓ-subgroups $G(g)$ and $G(h)$ are equal.

Now suppose that G is an ℓ-subgroup of H, and for each $h \in H^+$ there exists $g \in G^+$ so that h and g are a-equivalent. We then say that H is an *archimedean extension* (or *a-extension*) of G. If ℓ-group H admits no a-extensions, we call it *a-closed*; if an a-extension H of G is a-closed, then H is an *a-closure* of G. Thus Hölder's theorem asserts that each archimedean o-group admits a unique a-closure, namely the group of real numbers.

Because $\mathcal{C}(G)$ is a complete algebraic lattice with compact elements $G(g)$ (Theorem 1.2.2 and Proposition 1.2.3), it is a straightforward matter to verify that if H is an a-extension of G, then $\mathcal{C}(H)$ and $\mathcal{C}(G)$ are isomorphic as lattices, under the maps defined as follows: For $C \in \mathcal{C}(G)$ let $\Phi(C)$ be the smallest convex ℓ-subgroup of H containing C; for $D \in \mathcal{C}(H)$, let $\Psi(D) = D \cap G$. We summarize this information in the following Proposition:

Proposition 3.3 An ℓ-group H is an a-extension of ℓ-group G exactly when $\mathcal{C}(H)$ and $\mathcal{C}(G)$ are lattice isomorphic under the inverse maps Φ and Ψ.

We can easily reduce the question of existence of a-closures for a given ℓ-group to cardinality considerations by the following argument. If $\{G_\alpha : \alpha \in A\}$ is a chain of a-extensions of an ℓ-group G, and $G \subseteq G_\alpha \subseteq G_\beta$, whenever $\alpha < \beta$, it is easily seen that $\bigcup\{G_\alpha\}$ is also an a-extension of G. This means that if there exists a cardinality bound on a-extensions of a given ℓ-group, then by Zorn's lemma a-closures must exist. An important case where this argument applies is the class of abelian ℓ-groups. For if H is an abelian a-extension of an ℓ-group G, then $\mathcal{C}(G)$ and $\mathcal{C}(H)$ are isomorphic as ordered sets, and so by the Conrad-Harvey-Holland embedding theorem, H can be ℓ-embedded into $V(\Gamma(G), \mathbf{R})$; consequently the cardinality of the latter named group serves as the necessary bound. Thus each abelian ℓ-group admits an extension which is a-closed in the class of abelian ℓ-groups; actually, such ℓ-groups are a-closed in the class of all ℓ-groups (see Wolfenstein [67]).

The argument in the preceding paragraph does not guarantee, however, that any abelian a-closure of an abelian ℓ-group can be found inside the Hahn group of a given Conrad-Harvey-Holland embedding. To accomplish this requires a refinement of the proof of the Conrad-Harvey-Holland embedding theorem; namely, it is not necessary to use the entire root system of regular subgroups. The appropriate subsets are defined as follows:

A subset $\Delta \subseteq \Gamma(G)$ is *plenary* if $\cap\Delta = \{0\}$ and Δ is a *dual ideal* in $\Gamma(G)$; that is, if $\delta \in \Delta$, $\gamma \in \Gamma(G)$ and $\gamma > \delta$, then $\gamma \in \Delta$. We note that this is equivalent to the following:

(a) each $0 \neq g \in G$ has a value in Δ, and

(b) if $g \in G \backslash G^*_\gamma$ where $\delta \in \Delta$, then g has a value in Δ that exceeds δ.

The crucial example of a plenary subset is as follows: Let Δ be any root system, and for each $\delta \in \Delta$ let

$$G_\delta = \{g \in V(\Delta, \mathbf{R}) : g_\gamma = 0\ ,\ \text{for all } \gamma > \delta\}.$$

The set of all such G_δ's is a root system of regular subgroups of $V(\Delta, \mathbf{R})$ order isomorphic to Δ. Furthermore, it is a plenary subset of $\Gamma(V(\Delta, \mathbf{R}))$; obviously then we need not use the entire root system to obtain a Conrad-Harvey-Holland embedding. In fact, it is a straightforward matter to verify that for any abelian ℓ-group G, there exists a Conrad-Harvey-Holland embedding into $V(\Delta, \mathbf{R})$ where Δ is any plenary subset of $\Gamma(G)$.

We can now prove the following theorem:

Theorem 3.4 Let G be an abelian ℓ-group with Conrad-Harvey-Holland embedding into $V(\Delta, \mathbf{R})$, for some plenary subset Δ of $\Gamma(G)$, and let H be an abelian a-extension of G. Then H is ℓ-isomorphic to an ℓ-subgroup of $V(\Delta, \mathbf{R})$ containing G. In particular ℓ-groups of the form $V(\Delta, \mathbf{R})$ admit no abelian a-extensions.

Proof: Suppose that G is an abelian ℓ-group and H is a proper a-extension of G. Following our usual convention, let Δ denote both a plenary subset of the regular subgroups of G and an index set for that plenary subset. Since regular subgroups are distinguishable in the lattice $\mathcal{C}(G)$, which is lattice isomorphic to $\mathcal{C}(H)$, we can use this same index set for the corresponding plenary subset of the regular subgroups of H, and $G_\delta = G \cap H_\delta$.

Now by the Conrad-Harvey-Holland representation theorem we can ℓ-embed both G and H into $V(\Delta, \mathbf{R})$. We would like to define a map σ from $V(\Delta, \mathbf{R})$ onto itself so that the following diagram commutes (where ι, α and β are the obvious maps):

$$\begin{array}{ccc} H & \xrightarrow{\alpha} & V(\Delta,\mathbf{R}) \\ \uparrow{\scriptstyle \iota} & & \uparrow{\scriptstyle \sigma} \\ G & \xrightarrow{\beta} & V(\Delta,\mathbf{R}) \end{array}$$

In order to define this map, we must look first at each component of $V(\Delta, \mathbf{R})$. For each $\delta \in \Delta$, we have

$$G_\delta^*/G_\delta = G_\delta^*/(G_\delta * \cap H_\delta) \cong (G_\delta^* + H_\delta)/H_\delta \subseteq H_\delta^*/H_\delta;$$

call the map given here τ_δ. Furthermore, we have maps α_δ and β_δ injecting G_δ^*/G_δ and H_δ^*/H_δ into the real numbers $\mathbf{R}$. Thus $\sigma_\delta = (\beta_\delta)^{-1}\tau_\delta\sigma_\delta$ is an o-monomorphism from a subgroup of $\mathbf{R}$ to a subgroup of $\mathbf{R}$ and as such is just multiplication by some positive real number. (For suppose f is a non-zero o-monomorphism and $f(a_1)/f(a_2) \neq a_1/a_2$, say $f(a_1)/f(a_2) < a_1/a_2$. Then there exists two integers m and n such that $f(a_1)/f(a_2) < m/n < a_1/a_2$. We may assume that n and $a_2 > 0$ hence $ma_2 < na_1$ and $mf(a_2) > nf(a_1)$ which is a contradiction. This fact is due to Hion [54].) Consequently, we may as well suppose that σ_δ is an o-isomorphism from $\mathbf{R}$ onto itself.

We can now define σ as follows: for $v \in V(\Delta, \mathbf{R})$, let $\sigma(v)_\delta = \sigma_\delta(v_\delta)$. It now follows from Lemma 3.1 that σ is an ℓ-monomorphism; it is easily checked that σ is precisely the map which makes the diagram above commute. Furthermore, since the maps σ_δ are onto each component it is evident that σ is in fact onto. //

Unfortunately however, such abelian a-closures need not be unique (see E28). Furthermore, although they are in fact a-closures in the class of all ℓ-groups, not all a-closures of an abelian ℓ-group need be abelian (see E29). Now Khuon [70a] and McCleary (unpublished) have demonstrated the existence of a cardinality bound on a-extensions of an arbitrary ℓ-group, and so it follows that any ℓ-group admits a-closures (thus generalizing Conrad's result [54] that any (non-abelian) o-group admits an a-closure); needless to say, these a-closures need not be unique.

Conrad and Bleier [73],[75] made a further attempt to generalize Hahn's existence and uniqueness of a-closures. They considered extensions which preserve the lattice $\mathcal{K}(G)$ of closed convex ℓ-subgroups; such extensions are called a*-extensions. This is a natural generalization of the totally ordered case because there all convex subgroups are closed. Furthermore, because an ℓ-group is archimedean exactly if each closed convex ℓ-subgroup is a polar (Bigard [69a]), the concept of a*-extension in this case coincides with archimedean essential extension (see section 8.2 for a discussion of essential extensions of archimedean ℓ-groups). Ball [75] proved the existence of a*-closures for arbitrary ℓ-groups; however, they are usually not unique (see E29).

Chapter 4: *Representable and Normal-valued ℓ-groups*

Section 4.1: *The Lorenzen representation for ℓ-groups*

Clifford [40] and Lorenzen [49a] first observed that an abelian ℓ-group could be represented as a subdirect product of totally ordered groups; soon thereafter Lorenzen [49b] provided a characterization of those ℓ-groups admitting such a representation.

We shall call such ℓ-groups *representable*. The practical significance of such a representation is that it often enables one to apply techniques arising in both o-group and abelian ℓ-group theory. Furthermore, as we shall discover in Chapter 7, the class of representable ℓ-groups has an important role to play in the theory of varieties of ℓ-groups.

We say that G is a *subdirect product* of the ℓ-groups T_λ if G can be ℓ-embedded into the cardinal product ΠT_λ in such a way that each projection map is onto. An ℓ-group which admits no nontrivial subdirect product representation is said to be *subdirectly irreducible*; this is clearly equivalent to asserting that the ℓ-group has a minimum nontrivial ℓ-ideal.

Note that if G is an o-group and $a \neq b$ are elements of G, then $na \neq nb$ for all $n \neq 0$. Consequently $na = nb$ implies $a = b$; we say (speaking multiplicatively) that G satisfies the *unique extraction of roots*. It follows immediately that all representable ℓ-groups enjoy this property. This is not the case in general; for instance $A(\mathbf{R})$, the order-preserving permutations of the real line, does not have this property (see E51).

Clearly, an ℓ-group is a subdirect product of o-groups if and only if the ℓ-group possesses a collection of normal prime subgroups whose intersection is zero. Consequently, we observe that representability is a generalization of the abelian property. Note also that a subdirectly irreducible representable ℓ-group is obviously an o-group.

The following theorem gives several other characterizations of representable ℓ-groups:

Theorem 4.1.1 The following are equivalent for an ℓ-group G:

(a) G is representable.

(b) For all $a, b \in G$, $2(a \wedge b) = 2a \wedge 2b$.

(c) For all $a, b \in G$, $a \wedge (-b - a + b) \leq 0$.

(d) Each polar subgroup is normal.

(e) Each minimal prime subgroup is normal.

(f) For each $a \in G$, $a > 0$, $a \wedge (-b + a + b) > 0$, for all $b \in G$.

Proof: (a) ⇒ (b). Since a representable ℓ-group is a subdirect product of o-groups, we need only observe that the equality in (b) obviously holds in the totally ordered case.

(b) ⇒ (c). It follows from (b) that $2a \wedge 2b \leq a + b$. Applying this for $a = h + g$ and $b = h$, we have

$$((h+g)+(h+g)) \wedge (h+h) \leq h+g+h \ ; \text{ or,}$$

$$(g+h+g) \wedge h \leq g+h \ ; \text{ or,}$$

$$(h+g) \wedge (-g+h) \leq h \ ; \text{ or,}$$

$$g \wedge (-h-g+h) \leq 0.$$

(c) ⇒ (d). Suppose that $a \wedge b = 0$. We must show that $a \wedge (-c + b + c) = 0$. But because $a \wedge b = 0$, $a = (a - b) \vee 0$ and $b = (b - a) \vee 0$. Thus,

$$\begin{aligned} a \wedge (-c + b + c) &= ((a - b) \vee 0) \wedge (-c + ((b - a) \vee 0) + c) \\ &= ((a - b) \vee 0) \wedge ((-c + b - a + c) \vee 0) \\ &= ((a - b) \wedge (-c + b - a + c)) \vee 0. \end{aligned}$$

But this latter expression is 0, by (c) applied to $g = a - b$.

(d) ⇒ (e). Since each minimal prime is a union of polars (Theorem 1.2.11) this is obvious.

(e) ⇒ (a). Since each value contains a minimal prime it is obvious that the set of minimal primes has intersection zero. But then we have a collection of normal primes whose intersection is zero.

(a) ⇒ (f). Since each o-group satisfies (f) then so does each representable ℓ-group.

(f) ⇒ (d). Let $a \wedge b = 0$ then $a \wedge (-x + b + x) \wedge (x + a - x) \wedge b = 0$. Hence $q \wedge (x + q - x) = 0$ where $q = a \wedge (-x + b + x)$ and so $0 = q = a \wedge (-x + b + x)$. //

Notice that by (b) the representable ℓ-groups can be defined in terms of an equation. This becomes important when we consider varieties of ℓ-groups in Chapter 7.

Corollary 4.1.2 Each abelian ℓ-group is representable.

The representation theorem above thus looks at G in terms of the totally ordered sets G/P for certain primes P. The idea of doing this in the absence of normality leads in the next chapter to Holland's representation for any lattice-ordered group.

In the previous chapter we saw that every abelian ℓ-group could be embedded in a divisible ℓ-group. The natural approach to proving the corresponding theorem for representable ℓ-groups would be to embed such an ℓ-group into a product of o-groups, and then embed each o-group into a divisible o-group. Unfortunately, this latter problem (first posed by B. H. Neumann) has proved to be difficult; in the next chapter we shall see that Holland's representation theorem enables us to always embed an ℓ-group into a divisible ℓ-group.

Given a Lorenzen representation

$$G \longrightarrow \prod \{T_\lambda : \lambda \in \Lambda\}$$

of a representable ℓ-group, notice that we can interpret information about the polars of G in terms of Λ, as follows. For $X \subseteq G$, let $spt(X) = \bigcup\{spt(f) : f \in X\}$. Then, the map $P \to spt(P)$ from the polars $\mathcal{P}(G)$ to subsets of Λ is one-to-one (since the representation is faithful) and intersection-preserving. Consequently, $\{spt(P) : P \in \mathcal{P}(G)\}$ can be made into a Boolean algebra under inclusion, which is isomorphic to $\mathcal{P}(G)$. Note that the join is generally not set-theoretic union.

Two particularly nice examples of this idea are the following. If X is a Tychonoff space and G is $C(X)$, the ℓ-group of continous real-valued functions on X, we obtain exactly the regularly open subsets of X in this way. Indeed, the map $K \mapsto K''$ for $K \subseteq \Lambda$ is a closure operator which will always induce a topology on Λ. Of particular interest is the case in which that topology is discrete. This implies the existence of elements of G whose supports consist of the singletons of Λ; this implies that the Boolean algebra of polars must be atomic. In this latter condition (which can hold even if G is not representable), we say that G has a *basis*; more concretely, G possesses a maximal pairwise disjoint set of elements g_λ, where $G(g_\lambda)$ is an o-group. Such ℓ-groups have been much studied [C],[BKW].

We close this section by mentioning an interesting result due to H. Hollister [78] and Kopytov [82], generalizing the fact that each abelian ℓ-group is representable. We will not include the proof here:

Theorem 4.1.3 Every ℓ-group which is nilpotent as a group is representable.

Actually, since Hollister's and Kopytov's result Reilly [83] has shown that every nilpotent ℓ-group is weakly abelian, which is a condition that is considerably stronger than representability; see the discussion for the variety of weakly-abelian ℓ-groups, $\mathcal{W}$, in the appendix.

Section 4.2: *Normal-valued ℓ-groups*

We shall now digress from our program of examining representation theorems to consider an important class of ℓ-groups which contains the class of representable ℓ-groups; these are the normal-valued ℓ-groups. Read [74] has obtained a representation theory for such groups as a special case of Holland's representation theorem; we shall describe his results in Chapter 5. In fact, Read's results can be rephrased in terms of wreath products of the real numbers (see Theorem 7.1.10), thus providing a reasonably concrete representation for normal-valued ℓ-groups. In Chapter 10 we shall discuss how this theorem can be viewed as a generalization of the Conrad-Harvey-Holland representation theorem for abelian ℓ-groups.

Let P be a regular subgroup of an ℓ-group G and let P^* be its cover. Then P is said to be a *normal value* if P is normal in P^*. That is, P is a normal value if for all $x \in P^*$, $-x + P + x = P$. We say that G is *normal-valued* if all of its regular subgroups are normal values.

Not all ℓ-groups are normal-valued; $A(\mathbf{R})$ provides an example (see E51). Obviously, all abelian ℓ-groups are normal-valued. Furthermore, each representable ℓ-group is normal-valued (and so in particular each o-group is). To see this, recall from Theorem 4.1.1 that in a representable ℓ-group each minimal prime subgroup is normal. Let P be any regular subgroup of G. Now, P contains a minimal prime M, which is normal. Consequently, M is contained in all conjugates of P, and so the set of all conjugates of P forms a chain. Choose $g \in P^* \backslash P$. We know that P and $-g + P + g$ are comparable; we wish to show that they are equal. If not, and $P \subset -g + P + g$, then $P^* \subseteq -g + P + g$, the cover of P, and so $g \in -g + P + g$. But then $g \in P$, which is impossible. If $-g + P + g \subset P$, then $P \subset g + P - g$, which similarly leads to a contradiction. Therefore, $-g + P + g = P$. That is, P is a normal value, and so the representable ℓ-group G is normal-valued. This fact was first observed by Richard Byrd [66].

We need the following three lemmas before we can provide our characterizations of normal-valued ℓ-groups:

Lemma 4.2.1 Let G be an ℓ-group and P an ℓ-subgroup. If $x \in G$ with $P + x > P$ then for all $z \in P$, $P - x + z + x < P + x$.

Proof: If $P + x > P$, then $P > P - x$, and so $P = P + z > P - x + z$. Thus, $P + x > P - x + z + x$, as claimed. //

Lemma 4.2.2 Let P be a regular ℓ-subgroup of the ℓ-group G. Then the following are equivalent:

(a) P is a normal value.

(b) For all $x \in P^*$ and for all $y \in G^+ \backslash P$, there exists an integer n for which $P + x \leq P + ny$.

Proof: (a) $\Rightarrow$ (b). Suppose that P is a normal value for the ℓ-group G, and $x \in P^*$. If $y \in G^+ \backslash P^*$ it is always the case that $P + x < P + y$. For the case where $y \in P^*$ consider that P^*/P is a subgroup of the reals.

(b) $\Rightarrow$ (a). Let $x \in P^*$ and $P_1 = -x + P + x$; we wish to show that $P_1 = P$. Assume $x > 0$. Conjugation preserves inclusion and so both P and P_1 are maximal ℓ-subgroups of P^*. Hence, one cannot strictly contain the other. So, $P_1 \neq P$, and there exists $z \in P^+$ such that $-x+z+x \notin P$. But condition (b) implies that $P+x \leq P+n(-x+z+x) = P-x+nz+x$ for some n , and this contradicts Lemma 4.2.1. If we assume $x < 0$, we arrive at a similar contradiction. Since in general $x = x^+ - x^-$, it follows that $P = -x + P + x$.//

Lemma 4.2.3 The class of normal-valued ℓ-groups is closed under convex ℓ-subgroups.

Proof: Suppose G is a normal-valued ℓ-group and H is a convex ℓ-subgroup. Let A be a value in H of $h \in H$, with A^* its cover. Choose a value P of h in G which contains A. Then $P \cap H = A$ and $P^* \cap H \supseteq A^*$. But P is normal in P^* and so $P \cap H$ is normal in $P^* \cap H$; thus, A is normal in A^*. //

The following theorem gives important characterizations of normal-valued ℓ-groups. It is due to Wolfenstein [68a],[68b]; it is interesting to note that Birkhoff [42] in his original paper on ℓ-groups had asked whether condition (b) was satisfied by all ℓ-groups. Condition (c) will be particularly important when we consider varieties of ℓ-groups in Chapter 7. We first introduce a bit of notation and terminology. For an ℓ-group G and $a, b \in G^+$ we say that a is *infinitely larger* than b and write $b \ll a$ if $nb \leq a$ for all n.

Theorem 4.2.4 For an ℓ-group G the following are equivalent:

(a) G is normal-valued.

(b) For all $a, b \in G, -a - b + a + b \ll |a| \vee |b|$.

(c) For all $a, b \in G^+$, $a + b \leq 2b + 2a$.

(d) For all convex ℓ-subgroups A and B, $A + B = B + A = A \vee B$.

Proof: (a) $\Rightarrow$ (b). Assume that $a, b \in G^+$. By Lemma 4.2.3 we may as well assume that $G = G(|a| \vee |b|)$. Let H be the intersection of all values of $|a| \vee |b|$; that is, H is the intersection of all maximal convex ℓ-subgroups of G. Now, H is normal and G/H is a subdirect product of subgroups of the reals and as such is abelian. Then $-a - b + a + b$, as an element of the commutator subgroup of G, belongs to H.

We wish to show that $-a - b + a + b \ll |a| \vee |b|$; we shall in fact show that if $h \in H^+$, then $h \ll |a| \vee |b|$. If not, then $nh \not\leq |a| \vee |b|$ for some positive integer n. Let $x = nh - |a| \vee |b|$; we have that $x^+ > 0$. Choose V, a value of x^+; then $V \subseteq M$, a value of $|a| \vee |b|$. Now $x^- \in V$, since V is prime and $x^+ \notin V$. But $h \in H \subseteq M$, and so $x = nh - |a| \vee |b| \notin M$ (else, $|a| \vee |b| \in M$). Thus, $x^+ \notin M$. But then

$$M + nh - |a| \vee |b| = M + x = M + x^+ > M$$

and so

$$M + nh > M + |a| \vee |b| > M.$$

But then $h \notin M$, a contradiction; so, $h \ll |a| \vee |b|$.

(b) $\Rightarrow$ (c). If (b) is true, then for $a, b \in G^+$,

$$-b + a + b - a \leq |-b| \vee |a| = b \vee a \leq b + a.$$

Hence, $a + b \leq 2b + 2a$.

(c) $\Rightarrow$ (d). Suppose that A and B are convex ℓ-subgroups of G. We shall first show that the set $A + B$ is convex. For that purpose, suppose that $a_1 + b_1 \leq x \leq a_2 + b_2$, where $a_i \in A$ and $b_i \in B$, for $i = 1, 2$. Then

$$0 \leq -a_1 + x - b_1 \leq -a_1 + a_2 + b_2 - b_1 \leq |b_2 - b_1| + |-a_1 + a_2| + |b_2 - b_1|,$$

by (1.1.3). By the Reisz decomposition property (Theorem 1.1.2) we have that

$$-a_1 + x - b_1 = b_4 + a_3 + b_3 \text{ where } 0 \leq a_3 \leq |-a_1 + a_2| \text{ and } 0 \leq b_3, b_4 \leq |b_2 - b_1|.$$

But by (c), we have $b_4 + a_3 + b_3 \leq 2a_3 + 2b_4 + b_3$. By again applying the Reisz decomposition property, $-a_1 + x - b_1 = a_4 + b_5$. But then $x = (a_1 + a_4) + (b_5 + b_1)$, and so is an element of $A + B$.

Next, we shall show that $A + B = B + A$, and consequently $A + B$ is a subgroup of G. Toward that end, suppose that $g = a + b \in A + B$. Now, $|g| = |a + b| \leq |a| + |b| + |a| \leq 2|b| + 2|a| + |a| = 2|b| + 3|a|$. Similarly we also have that $|g| \leq 3|a| + 2|b|$, hence, $|g| \geq -2|b| - 3|a|$. Since $g \leq |g|$, $0 \leq 2|b| + g + 3|a| \leq 2|b| + |g| + 3|a| \leq 4|b| + 6|a|$. But then by the Reisz decomposition property we can express $2|b| + g + 3|a|$ as an element of $B + A$, and consequently $g \in B + A$. The other inclusion follows similarly.

Thus $A + B$ is a convex subgroup, and since it is clearly upper directed, it follows that $A + B$ is a convex ℓ-subgroup and hence equals $A \vee B$.

(d) $\Rightarrow$ (a). Suppose that condition (d) holds, let $x \in G^+$, and let V be a value of x. Choose $y \in G^+ \backslash V$; then we know that $V + G(y)$ is a convex ℓ-subgroup, which consequently contains V's cover V^*. But then $x = w + z$ with $z \in G(y)$ and $w \in V$. Then there exists n with $|z| \leq ny$, and so $V + x = V + z \leq V + ny$. But then, by Lemma 4.2.2 , V is a normal value. //

We knew already that the convex ℓ-subgroup C generated by two convex ℓ-subgroups A and B is just the subgroup generated by them (see Theorem 1.2.2); note that by (d), in the normal-valued case each element in C can be written as sums of the form $a + b$ (or $b + a$) , rather than as arbitrarily long sums of elements from A and B.

As promised, we now offer an alternative proof of the fact the every archimedean ℓ-group is abelian (Theorem 2.2). Recall that we have shown that if G is archimedean and $g \in G$ then g' is normal (Lemma 2.1). Hence G is representable by Theorem 4.1.1 and we may assume that G is an o-group. By part (b) of 4.2.4 we have that $-a - b + a + b \ll |a| \vee |b|$ but G is archimedean and so $-a - b + a + b = 0$.

Chapter 5: *Holland's Embedding Theorem*

Recall example 1.1.15: Let T be a totally ordered set and $A(T)$ the ℓ-group of order-preserving permutations of T, where $\alpha \in A(T)$ is positive if $t\alpha \geq t$ for all $t \in T$. Birkhoff [B] asked what ℓ-groups can be constructed in this manner. Holland [63] gave a partial answer to this question and in the process provided a new perspective from which ℓ-groups can be studied. (Bigard, Keimel and Wolfenstein [BKW] refer to this as "l'école américaine".) Holland's main result is that every ℓ-group is ℓ-isomorphic to an ℓ-subgroup of the ℓ-group of order-preserving permutations of some totally ordered set. In this chapter we will derive Holland's theorem, along with some immediate consequences. For a more complete study of ℓ-groups viewed in this manner see [G]. The reader should note that after Holland's theorem is proved we will employ multiplicative notation extensively.

Lemma 5.1 If P is a prime subgroup of an ℓ-group G, then each $g \in G$ induces an order-preserving permutation $\beta(g, P)$ of G/P (the right cosets of P), defined by $(P + x)\beta(g, P) = P + x + g$.

The proof is straightforward.

We need the following definition. Let T be a totally ordered set; an ℓ-subgroup H of $A(T)$ is *transitive* (or *acts transitively* on T) if for all $t, s \in T$, there exists $h \in H$ such that $th = s$.

Lemma 5.2 If P is a prime subgroup of an ℓ-group G then the mapping

$$\alpha(P) : G \to A(G/P) \text{ defined by } g\alpha(P) = \beta(g, P)$$

is an ℓ-homomorphism of G onto a transitive ℓ-subgroup of $A(G/P)$.

Proof: One of the things we need to show is that $(g \vee h)\alpha(P) = \beta(g, P) \vee \beta(h, P)$. That is, for any right coset $P + x$,

$$P + x + (g \vee h) = (P + x + g) \vee (P + x + h).$$

But this follows from the proof of Theorem 1.2.1(b). The rest is similarly easy to verify. //

We call the homomorphism constructed in Lemma 5.2 a *right regular representation* of G in $A(G/P)$.

Theorem 5.3 Every ℓ-group is ℓ-isomorphic to a subdirect product of ℓ-groups, each of which is a transitive ℓ-subgroup of the ℓ-group of order-preserving permutations of a totally ordered set.

Proof: For each $0 \neq g \in G$ let $V(g)$ be a value of g (which exists by Proposition 1.2.9). Now $V(g)$ is prime (Theorem 1.2.10) and so by Lemma 5.2 the right regular representation $\alpha(V(g)) : G \to A(G/V(g))$ is an ℓ-homomorphism of G onto a transitive ℓ-subgroup of $A(G/V(g))$. Call this ℓ-subgroup $K(g)$. Now,

$$\theta : G \longrightarrow \prod\{K(g) : 0 \neq g \in G\}$$

is an ℓ-homomorphism, where $(h\theta)_g = h\alpha(V(g))$. Furthermore, $Ker(\theta)$ is trivial because any $0 \neq g \in Ker(\alpha(V(g)) = V(g)$. Hence, θ is an embedding. //

Theorem 5.4 (Main Embedding Theorem) Any ℓ-group is ℓ-isomorphic to an ℓ-subgroup of the ℓ-group of order-preserving permutations of a totally ordered set.

Proof: From Theorem 5.3 there is an embedding

$$\theta : G \longrightarrow \prod\{A(G/V(g)) : 0 \neq g \in G\}.$$

Choose a total order of the index set for this product (that is, $G\backslash\{0\}$). Now totally order $T = \bigcup\{G/V(g) : 0 \neq g \in G\}$ lexicographically: that is, by setting $x \prec y$ if $x, y \in G/V(g)$ and $x < y$ as elements of $G/V(g)$; or if $x \in G/V(g)$ and $y \in G/V(h)$ where $g < h$ under the total ordering given above to $G\backslash\{0\}$. We now embed $\prod A(G/V(g))$ into $A(T)$ as follows: if $\phi \in \prod A(G/V(g))$, then ϕ induces an order-preserving permutation $\overline{\phi}$ on T by $x\overline{\phi} = x\phi_g$, where $x \in G/V(g)$ and ϕ_g is the gth component of ϕ. The details can by easily checked. //

Since we can now view the elements of an ℓ-group as permutations of some totally ordered set, we will express the group operation multiplicatively (with 1 denoting the identity) for the remainder of this chapter. Furthermore, we shall always denote by T a totally ordered set.

We next give a description of ℓ-groups for which an embedding in Theorem 5.3 can be chosen so that there is only one summand.

Theorem 5.5 An ℓ-group G is ℓ-isomorphic to a transitive ℓ-subgroup of $A(T)$ for some T if and only if there exists a prime subgroup C of G such that the only ℓ-ideal of G contained in C is $\{1\}$.

Proof: Suppose G is a transitive ℓ-subgroup of $A(T)$, for some T. Choose $x \in T$ and let $C = \{g \in G : xg = x\}$; then C is a convex ℓ-subgroup of G, as can be easily verified. Furthermore, C contains no ℓ-ideals of G. For if $1 \neq g \in C$ then there exists $y \in T$ with $yg \neq y$. Then, since G is transitive, there exists $f \in G$ such that $xf = y$. Hence $xfgf^{-1} = ygf^{-1} \neq yf^{-1} = x$, and so $fgf^{-1} \notin C$.

Next we show that C is prime. Otherwise by Theorem 1.2.10 there exist $a, b \notin C$ with $a \wedge b = 1$. In other words $xa \neq x \neq xb$, and yet $x = x1 = x(a \wedge b) = xa \wedge xb$. But this is impossible since T is totally ordered.

Conversely suppose that C is such a subgroup of G. Then, by Lemma 5.2 the mapping $\alpha(C)$ is an ℓ-homomorphism of G onto a transitive ℓ-subgroup of $A(G/C)$. If $g \in Ker(\alpha(C))$ then $Cg = C$ and so $Ker(\alpha(C)) \subset C$. But the kernel is an ℓ-ideal and so is $\{1\}$; thus, $\alpha(C)$ is one-to-one. //

The convex ℓ-subgroup C described in the previous theorem is called a *representing* subgroup of G. Notice that $\{1\}$ is a representing subgroup if and only if $\{1\}$ is prime if and only if G is an o-group.

Corollary 5.6 Any subdirectly irreducible ℓ-group is ℓ-isomorphic to a transitive ℓ-subgroup of $A(T)$ for some T.

Proof: Let K be the minimum proper ℓ-ideal of a subdirectly irreducible ℓ-group G. Choose $1 \neq g \in K$ and let $V(g)$ be a value of g. Then $V(g)$ is a representing subgroup of G. //

Corollary 5.7 An abelian ℓ-group which is a transitive ℓ-subgroup of $A(T)$ for some T is totally ordered.

Proof: Let G be such an abelian ℓ-group and let C be its representing subgroup. Then C is an ℓ-ideal, and so $C = \{1\}$, since C contains only $\{1\}$ as an ℓ-ideal. Hence G is totally ordered. //

An interesting related fact due to P. M. Cohn [57] is that $A(T)$ is a totally orderable group if and only if $A(T)$ is abelian. The proof is beyond the scope of this text.

The applications of Holland's embedding theorem to the theory of ℓ-groups are manifold; we will mention only a few here.

The first of these applications provides us with a representation for normal-valued ℓ-groups, due to J. A. Read [74]. To describe this, we need a bit of terminology. Let $g \in A(T)$ and $t \in T$. The convex subset of T swept out by images of t under g is denoted by $\Delta(t, g)$. That is,

$$\Delta(t,g) = \{s \in T : t|g|^n \leq s \leq t|g|^{n+1}, \text{ for } n \in \mathbf{Z}^+\}.$$

We say that ℓ-subgroup G of $A(T)$ is *overlapping* if there are $t \in T$ and $f, g \in G$ such that $\Delta(t, g)$ and $\Delta(t, f)$ are incomparable with respect to inclusion. Read has shown that an ℓ-subgroup of $A(T)$ is normal-valued if and only if it is non-overlapping; we will discuss this result further in Chapter 10. For a good treatment of this and other results about normal-valued ℓ-groups from the permutation group point of view, see Chapter 10 of [G].

An important property which $A(T)$ possesses is lateral completeness. An ℓ-group is *laterally complete* if any collection of pairwise disjoint elements has a least upper bound. To see that this is true for $A(T)$, we first observe that elements of $A(T)$ are disjoint in the ℓ-group sense exactly if their supports are pairwise disjoint. Consequently if $\{g_\alpha\}$ is a collection of pairwise disjoint elements from $A(T)$, we may define $g \in A(T)$, which must be their least upper bound, as follows:

$$xg = \begin{cases} x & \text{if } xg_\alpha = x \text{ for all } g_\alpha \\ xg_\beta & \text{if } xg_\beta \neq x \text{ for some } g_\beta \end{cases}$$

Thus, it follows from Holland's embedding theorem (5.4) that every ℓ-group can be ℓ-embedded in a laterally complete ℓ-group. Much work has been done to construct the lateral completion of an ℓ-group G (a laterally complete ℓ-group containing G which is minimal in this regard). See Chapter 9 for an extended discussion of this problem.

Another important consequence of Holland's embedding theorem is that every ℓ-group can be ℓ-embedded into a divisible ℓ-group, thus providing the ℓ-group analogue of Neumann's [43] famous group-theoretic result. (Note that it is unknown whether any o-group can be o-embedded into a divisible o-group; see the discussion of this in Chapter 4 and E57).

To discuss this result requires some terminology. An ℓ-subgroup G of $A(T)$ acts *o-2 transitively* on T if for all $s_1, s_2, t_1, t_2 \in T$ with $s_1 < s_2$ and $t_1 < t_2$, there exists $g \in G$ with $s_i g = t_i$, for $i = 1, 2$; we say that T is *o-2 homogeneous* in this case. The reader may easily verify that if T is an ordered field (for example the reals) then T is o-2 homogeneous.

Note that we could define *o-n transitivity* analogously; however,the following result shows that this is no real generalization (for ℓ-groups at least; see [G]). While not immediately important, we will need this theorem in Chapter 7:

Theorem 5.8 An o-2 transitive ℓ-subgroup of $A(T)$ is o-n transitive, for any positive integer n.

Proof: Let G be such an ℓ-group, and $s_1 < s_2 < s_3$, $t_1 < t_2 < t_3$ be elements of T. Let $f \in G$ with $s_1 f = t_1$ and $s_2 f = t_2$. If $s_3 f \geq t_3$ then choose $g \in G$ with $s_1 g = t_2$ and $s_3 g = t_3$. Then $s_i(f \wedge g) = t_i$ for $i = 1, 2, 3$. If $s_3 f < t_3$, choose $g \in G$ with $s_2 g = t_1$ and $s_3 g = t_3$. Then $s_i(f \wedge g) = t_i$ for $i = 1, 2, 3$. Thus G is o-3 transitive. Induction extends this to any positive integer n. //

We can now describe the proof that every ℓ-group can be ℓ-embedded into a divisible ℓ-group. Given any ℓ-group, we can ℓ-embed it into an $A(T)$, for some totally ordered set T. We can then embed the chain T into an o-2 homogeneous chain S,and each element of $A(T)$ can be extended to a member of $A(S)$ in a fashion which ℓ-embeds $A(T)$ into $A(S)$ (we will comment further on this step of the proof below). To complete the proof we need only check that $A(S)$ is divisible; this verification is straightforward and computational, and can be found in Holland [63] or [G] (Section 2.2).

Holland's original proof [63] that a chain can be o-embedded into an o-2 homogeneous chain as described above uses the theory of η_α-fields, and in particular depends on a set-theoretic assumption equivalent to the generalized continuum hypothesis. Fortunately, Weinberg [67] was able to avoid this restriction by the following means. Given a totally ordered set T, let $G(T)$ be the free abelian group on T, ordered lexicographically. Then consider the rational group algebra $\mathbf{Q}[G(T)]$; this can also be totally ordered lexicographically, and this order has a unique extension to its quotient field S. It is straightforward to verify that every order-preserving permutation of T can be lifted to one of S in such a way that ℓ-embeds $A(T)$ into $A(S)$. Since S is a field it is o-2 homogeneous and consequently $A(S)$ is divisible. For full details the reader should consult Weinberg's paper [67].

For our last application of Holland's theorem we will return to the "ℓ-groupability" question discussed in Section 1.1. We will now characterize group-theoretically those groups which appear as subgroups of an ℓ-group.

A group G is said to be *right-ordered* if it admits a total order which is preserved by the group operation on the right; that is, if $f, g, h \in G$ and $f \leq g$ then $fh \leq gh$. Obviously all o-groups are right-ordered, but the converse is false (see E2). Such groups were first considered by Conrad [59], Zaitzeva [58] and Cohn [57] and are covered in [MR].

In particular Conrad provided a group-theoretic characterization of those groups admitting a right order, similar to the Ohnishi [50] conditions describing those groups admitting an o-group structure. We shall not demonstrate this here, but shall accept that right-orderability is a group-theoretic property, and characterize subgroups of ℓ-groups in the following theorem:

Theorem 5.9 A group G is isomorphic to a subgroup of an ℓ-group if and only if G is right-orderable.

Proof: Suppose that G is a subgroup of an ℓ-group. By Holland's theorem we may as well assume that G is a subgroup of $A(T)$. Consequently we need only equip $A(T)$ with a right order. To do this well order T. For $f, g \in A(T)$, we say that $f \prec g$ exactly when $xf < xg$ (in the usual order on T), where x is the first element of T in its well-ordering at which $xf \neq xg$. It is easy to see that this is a total ordering which multiplication preserves on the right.

Conversely if G is a right-ordered group, we can clearly embed G into $A(G)$ via the right regular representation $g \mapsto \alpha(g)$, where $h\alpha(g) = hg$. Hence G is isomorphic to a subgroup of an ℓ-group. //

We can in fact refine this result to characterize those partially ordered groups which can be obtained as subgroups of ℓ-groups with the induced order. To do this consider again the method in the proof above of right-ordering $A(T)$. Clearly for each $t \in T$ we may choose a distinct well-ordering of T which makes t the first element. These well-orderings lead to distinct right orders on $A(T)$, and clearly the intersection of the positive cones of these orders is exactly $A(T)^+$. Because of Holland's embedding theorem we then have the following (due to Hollister [65]):

Proposition 5.10 The positive cone of an ℓ-group is an intersection of right orders for the group.

If G and H are partially ordered groups, G is a subgroup of H, and the order on G is induced from that on H, we call G a *po-subgroup* of H. A group homomorphism from G into H is called an *o-homomorphism* if it preserves order; if an o-homomorphism is one-to-one, and preserves order in both directions, we call it an *o-monomorphism*. Note that the image of an o-monomorphism is necessarily a po-subgroup. We can now characterize po-subgroups of ℓ-groups:

Proposition 5.11 A partially ordered group admits an o-monomorphism onto a po-subgroup of an ℓ-group exactly if its positive cone is an intersection of right orders.

Proof: Suppose that G is a partially ordered group and $G^+ = \cap P_\lambda$, where the P_λ's are right orders for G. Let G_λ be G equipped with the right order P_λ and let $g \mapsto g_\lambda$ be the right regular representation of G_λ in $A(G_\lambda)$. If $g \in G^+$ then $g_\lambda \in A(G_\lambda)^+$, for each λ. On the other hand, if each g_λ is positive then $1 \leq 1g_\lambda = g$ in G_λ and so $g \in \cap P_\lambda = G^+$. Thus the map

$$g \longmapsto (\ldots, g_\lambda, \ldots) \in \prod A(G_\lambda)$$

is an order-preserving embedding of G into an ℓ-group.

The converse follows immediately from Proposition 5.9. //

As we shall see, these results become important in the construction of free ℓ-groups, which we will do in the next chapter.

Chapter 6: *Free ℓ-groups*

Because the class of lattice-ordered groups is a variety, it follows from universal algebra that there exist free objects for the class [Co]; in fact, the same observation holds for any variety of lattice-ordered groups, and in particular for the variety of abelian ℓ-groups. (In the next chapter we shall consider varieties of ℓ-groups in some detail.) However it took considerable work by several mathematicians before a satisfactory construction of free ℓ-groups was obtained. Birkhoff [42] was the first to mention free ℓ-groups; he described explicitly the free ℓ-group on one generator. Baker, in his Ph.D. thesis (Harvard, 1966), and Weinberg [63], [65] first constructed the free abelian ℓ-group over a set of generators, and more generally, over a partially ordered abelian group (see also Henriksen and Isbell [62], whose work in a category of lattice-ordered rings predated Weinberg's). Work of Birkhoff [B], Baker [68] and Beynon [74] has led to a more intuitively satisfying description of the free abelian ℓ-group on a set of generators. Bernau [70] first constructed nonabelian free ℓ-groups, using an equivalence relation on the set of all non-empty finite subsets of the partially ordered group. Conrad [70a] then generalized Weinberg's construction to the nonabelian case, thus providing a more understandable representation for the free ℓ-group. Recently McCleary [85a], [85b] has used permutation group methods to considerably add to our understanding of arbitrary free ℓ-groups; we will briefly discuss his results at the end of this chapter.

We shall begin our discussion with a general defintion. Let S be an arbitrary set and π a one-to-one map from S into the ℓ-group F. Then (F, π) is the *free ℓ-group* over S if

(a) the ℓ-subgroup of F generated by $S\pi$ is F,

(b) if σ is a map of S into an ℓ-group H, then there exists an ℓ-homomorphism τ from F into H such that $\pi \circ \tau = \sigma$; that is, the following diagram commutes:

$$\begin{array}{ccc} S & \xrightarrow{\pi} & F \\ \sigma\searrow & & \swarrow\tau \\ & H & \end{array}$$

We say in this case that S is a *free set of generators* for F.

We can modify this definition in several ways. First, we can equip S with some structure by making it a partially ordered group. We then naturally insist that π be an o-monomorphism and σ range over all o-homomorphisms, thus providing us with the definition of the free ℓ-group (F, π) over a partially ordered group S. Note, however, that we then have no guarantee that such objects exist. We shall see later that the original definition can be viewed as a special case of this definition (Corollary 6.6).

Secondly, we can restrict our attention to only abelian ℓ-groups (or for that matter, to ℓ-groups belonging to any ℓ-group variety; see Chapter 7). We then obtain definitions for the free abelian ℓ-groups over arbitrary sets and over (abelian) partially ordered groups.

Suppose now that F is a free ℓ-group on the set S. We wish to inquire more carefully into what this means in terms of the generators themselves. Note that the discussion which follows applies equally well if F were a free abelian ℓ-group on a set S (except that we would use additive notation).

Since S generates F as an ℓ-group, each element of F can by Proposition 1.2.4 be put into the form

$$w = \bigvee_I \bigwedge_J \prod_K s_{ijk},$$

where I, J and K are finite index sets, and for each ijk, either s_{ijk} or s_{ijk}^{-1} is an element of S. We say in this case that w has been put into *normal form*. Unfortunately the normal form is not unique, and indeed, such expressions can well represent the identity even if not all the group elements $\prod s_{ijk}$ equal the identity. (For example, consider $(a \wedge a^{-1}) \vee aa^{-1} = 1$).

Suppose that w is a word obtained from the free generators of a free ℓ-group, and it represents the identity. If we replace each occurrence of each generator by a corresponding element from some arbitrary ℓ-group G, we obtain (after computation) an element in G which must also be the identity. For otherwise the obvious map from S into G would not have a lifting to F. (We shall express this in the following chapter by saying that an equation satisfied by a free ℓ-group is satisfied by all ℓ-groups.)

Note that we can now turn the preceding discussion around. Namely, suppose that S is a set of generators for an ℓ-group G and that the only words w obtained from the set S which represent the identity also represent the identity when the corresponding word is formed from a free set of generators (of the same cardinality) for a free ℓ-group F. This means that an ℓ-isomorphism exists from F onto G, and so S is a free set of generators for the (necessarily free) ℓ-group G. We can thus describe free ℓ-groups (and free abelian ℓ-groups) in terms of the "freeness" of the generators, rather than in terms of the mapping property of the definition.

We now wish to construct free ℓ-groups. We shall begin with the case of the free abelian ℓ-group over a set of generators. In order to do this, we must prove the following important theorem of Weinberg [63]; this is also vitally important for our study of varieties in the next chapter.

Theorem 6.1 The free abelian ℓ-group on a set of generators is a subdirect product of copies of the ordered group of integers; in particular, it is archimedean.

Proof: We first observe that each free abelian ℓ-group is a subdirect product of free abelian ℓ-groups on finitely many generators (this result holds in any class of algebras [Co]). Thus, to prove the theorem, we need only show that a free abelian ℓ-group on finitely many generators is a subdirect product of integers. Let F be such an ℓ-group, free on a finite set X. Since F is abelian we know that it is a subdirect product of a collection of abelian o-groups G_λ; as ℓ-homomorphic images of F, each G_λ is finitely generated. We will prove below that each such G_λ is then an ℓ-homomorphic image of a subdirect product S_λ of integers via a map ϕ_λ. But this will be sufficient, because by the freeness of F we can find an ℓ-homomorphism τ_λ from F into S_λ so that $\tau_\lambda \circ \sigma_\lambda$ is the projection of F onto G_λ; putting these together shows that F is a subdirect product of the S_λ's . Thus, it remains to show that every finitely generated abelian o-group is an ℓ-homomorphic image of a subdirect product of integers. This is the content of Lemma 6.2 below. //

Weinberg's proof [63] of the following lemma, which is needed to complete the proof of Theorem 6.1, involved the theory of η_1-fields. The proof we give here is due to Holland, Mekler and Reilly [86].

Lemma 6.2 Each finitely generated abelian o-group can be obtained as an ℓ-subgroup of an ℓ-homomorphic image of a countable product of integers.

Proof: Let G be a finitely generated abelian o-group. By Hahn's representation theorem for abelian o-groups, G can be o-embedded into a lexicographically ordered o-group ΠT_i, where each T_i is a finitely generated subgroup of the real numbers. We shall in fact prove the lemma for the full group ΠT_i. Let $\{t(i,j)\}$ be a finite set of generators for T_i; clearly, we can assume without loss of generality that $t(i,0) = 1$ and $t(i,j) > 0$, for all j. Now consider the cardinal product of integers ΠZ, indexed by the positive integers $\mathbf{N}$. Choose any non-principal ultrafilter $\mathcal{F}$ on $\mathbf{N}$ and let M be the ℓ-ideal which consists of all elements of the product whose supports are complementary to elements of $\mathcal{F}$. We shall embed ΠT_i into $\Pi\mathbf{Z}/M$ as follows. For each generator $t(i,j)$, we can choose a sequence of integers $\{n(i,j,k)\}$ such that

(1) $$\lim_{k\to\infty} \frac{n(i,j,k)}{n(i,0,k)} = t(i,j), \text{ and}$$

(2) $$\lim_{k\to\infty} \frac{n(i',j',k)}{n(i,j,k)} = \infty, \text{ if } i < i'.$$

It is then easy to see that the subgroup of $\Pi\mathbf{Z}/M$ generated by the images of these sequences is o-isomorphic to ΠT_i, as required. //

Note that Conrad [71b] has given an explicit description of the generators of a free abelian ℓ-group, as elements of a cardinal product of integers.

We are now ready to construct free abelian ℓ-groups, following the approach of Borkhoff [B], Baker [68] and Beynon [74]. Let α be any cardinal and let $\mathbf{R}^\alpha$ be the real vector space consisting of the (unrestricted) sum of α copies of $\mathbf{R}$. Our description of the free abelian ℓ-group on α generators will be as a subgroup of the additive group of real-valued functions on $\mathbf{R}^\alpha$.

To accomplish this we need the following definitions. A function $f : \mathbf{R}^\alpha \to \mathbf{R}$ is *homogeneous linear* if there exist real numbers $\{a_\gamma\}$, only finitely many of which are nonzero, such that

$$f(\vec{x}) = \sum a_\gamma x_\gamma;$$

if all the a_γ's are integers we say f *has integer coefficients*. This is of course well-defined because only finitely many of the a_γ's are nonzero. A function f is *piecewise homogeneous linear* if it is continuous, and there exist finitely many homgeneous linear functions $f_1, \ldots, f_k$

such that for all $\vec{x} \in \mathbf{R}^\alpha$, there exists i such that $f(\vec{x}) = f_i(\vec{x})$; we say that f *has integer coefficients* if each of the f_i's does. We denote by F_α the set of all piecewise homogeneous linear functions with integer coefficients; F_α is evidently an archimedean ℓ-group under addition and the pointwise order. We shall assert in the next theorem that F_α is the free abelian ℓ-group on α generators:

Theorem 6.3 The ℓ-group F_α of piecewise homogeneous linear functions with integer coefficients is the free abelian ℓ-group on α generators.

Proof: Denote by p_γ the projection map on the γth component; that is, for $\vec{x} = (\ldots x_\gamma \ldots)$, $p_\gamma(\vec{x}) = x_\gamma$. We wish to prove that $\{p_\gamma\}$ is a free set of generators for F_α. To show this we must check that $\{p_\gamma\}$ is free: that is, this set satisfies only equations satisfied by all abelian ℓ-groups. For this purpose, suppose that w is a word in normal form in n variables; denote by $\overline{w}$ the element of F_α which results when n of the functions p_{γ_i} are substituted in for the variables of w. Suppose now that this piecewise homogeneous linear function $\overline{w}$ is the zero function. Consider a set of integers $\{b_1, \ldots, b_n\}$, and obtain the element $\vec{b}$ of $\mathbf{R}^\alpha$ whose γ_i entries are the b_i's, and whose other coordinates are zero. Then $\overline{w}(\vec{b}) = 0$. But $\overline{w}(\vec{b})$ is exactly the result of replacing the n variables of w by the integers $b_1, \ldots, b_n$. Since every such substitution of integers into the word w thus gives 0, this means that the equation $w = 0$ is satisfied by the ℓ-group $\mathbf{Z}$. But then Theorem 6.1 implies that $w = 0$ is satisfied by all abelian ℓ-groups, and so $\{p_\gamma\}$ is a free set of generators for the ℓ-subgroup of F_α which they generate.

We must now show that $\{p_\gamma\}$ actually generates the entire ℓ-group F_α. We first observe that $\{p_\gamma\}$ clearly generates (group-theoretically) the subgroup of F_α consisting of all homogeneous linear functions with integer coefficients. It thus remains to show that each element of F_α can be expressed as a finite join of finite meets of homogeneous linear functions with integer coefficients.

Let $f \in F_\alpha$; then there exist $f_1, \ldots, f_k$, homogeneous linear functions with integer coefficients, so that for each $\vec{x} \in \mathbf{R}^\alpha$, $f(\vec{x}) = f_i(\vec{x})$, for some i. We may as well assume that the f_i's are distinct. We shall actually show that f equals a finite join of finite meets of the f_i's , for appropriately chosen subsets.

For each $i \neq j$, $i, j \leq k$, consider the subspace

$$H_{ij} = \{\vec{x} \in \mathbf{R}^\alpha : f_i(\vec{x}) = f_j(\vec{x})\};$$

this is clearly of dimension strictly less than α. Consider now $\bigcup_{i \neq j} H_{ij}$. If $\vec{x} \notin \bigcup H_{ij}$, then

$$f_{\sigma(1)}(\vec{x}) > \ldots > f_{\sigma(k)}(\vec{x}),$$

for some permutation $\sigma \in S_k$, the group of permutations of $\{1, 2, \ldots, k\}$. Thus, if we let

$$T_\sigma = \{\vec{x} \in \mathbf{R}^\alpha : f_{\sigma(1)}(\vec{x}) > \ldots > f_{\sigma(k)}(\vec{x})\},$$

we have that

$$\mathbf{R}^\alpha \setminus \bigcup H_{ij} = \bigcup \{T_\sigma : \sigma \in S_k\}.$$

Since each H_{ij} is a proper subspace, this means that $\mathbf{R}^\alpha$ is spanned by the set $\bigcup\{T_\sigma : \sigma \in S_k\}$. In particular, this means that for some permutations σ we have that $T_\sigma \neq \emptyset$; let Σ be the set of such. Note that each (nonempty) T_σ is *convex*, meaning here that if $\vec{x}, \vec{y} \in T$, then the line segment $[\vec{x}, \vec{y}] \subseteq T$.

Because $\bigcup T_\sigma$ is a spanning set, we can (for the most part) restrict our attention to the behavior of f on this set. In particular, we wish to observe that $f|T_\sigma = f_i$, for some i : for if $\vec{x}, \vec{y} \in T_\sigma$, and $f(\vec{x}) = f_i(\vec{x})$ while $f(\vec{y}) = f_j(\vec{y})$, for $i \neq j$, then by continuity there exists some point $\vec{z} \in [\vec{x}, \vec{y}]$ at which $f_i(\vec{z}) = f_h(\vec{z})$, for some $h \neq i$. But then $\vec{z} \in H_{ih} \cap T_\sigma = \emptyset$, which is a contradiction.

We are now ready to define the lattice word in the f_i's which we shall prove equals f. Given $\rho \in \Sigma$, there exists some j for which $f|T_\rho = f_{\rho(j)}$. Let $g_\rho = \bigwedge_{i=1}^{j} f_{\rho(i)}$. It is quite evident that $g_\rho = f$ on T_ρ . Consequently, $\bigvee_{\rho\in\Sigma} g_\rho \geq f$ on $\bigcup T_\rho$, and so on all of $\mathbf{R}^\alpha$. It thus remains to show that each $g_\rho \leq f$ in order to prove that

$$f = \bigvee_{\rho\in\Sigma} g_\rho = \bigvee \bigwedge f_{\rho(i)},$$

as required. We of course need only to check that $g_\rho \leq f$ on $\bigcup T_\sigma$.

To accomplish this, we shall suppose that $\vec{y} \in \bigcup T_\rho$, $f(\vec{y}) = f_{\rho(h)}(\vec{y})$ and that $f|T_\rho = f_{\rho(j)}$. If $j \leq h$, then

$$g_\rho(\vec{y}) = \bigwedge_{i=1}^{j} f_{\rho(i)}(\vec{y}) \leq f_{\rho(h)}(\vec{y}) = f(\vec{y}),$$

as required.

We shall thus assume that $h > j$. Choose $\vec{x} \in T_\rho$, and consider the graph of f on the line segment $[\vec{x}, \vec{y}]$. In particular, we wish to compare this graph to that of the linear function s on $[\vec{x}, \vec{y}]$ which agrees with f at $\vec{x}$ and $\vec{y}$. Since $\vec{x} \in T_\rho$ and $h > j$, we have $f(\vec{x}) = f_{\rho(j)}(x) > f_{\rho(h)}(x)$, while $f(\vec{y}) = f_{\rho(h)}(\vec{y})$. This means that $f(\vec{z}) < s(\vec{z})$, for some $\vec{z} \in [\vec{x}, \vec{y}]$. Let $\vec{w}$ be the point on the line segment $[\vec{x}, \vec{y}]$ farthest from x at which $f(\vec{w}) = f_{\rho(j)}(\vec{w})$; note that $\vec{w}$ may be $\vec{x}$, but can certainly not be $\vec{y}$. In the direction of $\vec{y}$ from $\vec{w}$ on $[\vec{x}, \vec{y}]$ it may be the case that f agrees with f_r for some r , and the graph of f is certainly below the graph of s there. This means (since f_r is linear) that $f_r(\vec{y}) < f(\vec{y})$, while $f_r(\vec{x}) \geq f(\vec{x})$. But then $r = \rho(m)$ for some integer m , and $m \leq j$. But then

$$g_\rho(\vec{y}) = \bigwedge_{i=1}^{j} f_{\rho(i)}(\vec{y}) \leq f_{\rho(m)}(\vec{y}) < f(\vec{y});$$

this is what we wished to prove. //

It was Birkhoff [B] and Baker [68] who first observed that the projections $\{p_\gamma\}$ satisfy no nontrivial abelian ℓ-group equations. Beynon [74] proved that the projections in fact generate F_α as an ℓ-group; Madden [85] has provided an alternate proof. This construction has been extended by Beynon to provide an elegant (and algorithmically effective) correspondence between finitely presented abelian ℓ-groups and closed polyhedral cones of $\mathbf{R}^n$. Glass and Madden [85] noted that this correspondence is algorithmically effective.

We now turn to the construction of arbitrary free ℓ-groups; we will describe Conrad's [70] construction. Suppose for a moment that the free ℓ-group F exists over some partially ordered group G. Observe that since $G\pi$ is a po-subgroup of the ℓ-group F it follows from Proposition 5.10 that the positive cone of $G\pi$ must be the intersection of right orders. We shall prove in the next theorem that the free ℓ-group over any such partially ordered group exists. The essential idea for the proof is that words in such a partially ordered group are distinguished in the ℓ-group of order-preserving permutations of it under some right ordering. The technical details of this fact will be verified in Lemma 6.5 whose proof follows the theorem itself.

Theorem 6.4 Let G be a partially ordered group whose positive cone is an intersection of right orders. Then there exists a unique free ℓ-group over G.

Proof: We shall leave to the reader the standard proof that such a free ℓ-group is unique, and proceed with its construction as follows: Suppose that $G^+ = \cap P_\lambda$, where $\{P_\lambda : \lambda \in \Lambda\}$ is the set of all right orders on G extending its given partial order. For each $\lambda \in \Lambda$, let G_λ be the right ordered group obtained when G is endowed with the right order P_λ. As in the proof of Proposition 5.11, we can then embed G into the ℓ-group $\prod A(G_\lambda)$ via the map π whose λth component is just the right regular representation of G in $A(G_\lambda)$. Let F be the ℓ-subgroup of $\prod A(G_\lambda)$ generated by $G\pi$.

To prove that the ℓ-group F constructed above is the free ℓ-group over G, we suppose that σ is an o-homomorphism of G into an ℓ-group H. We wish to lift σ to F. Now, an element of F can be written in the form $\vee_I \wedge_J a_{ij}\pi$, where I and J are finite sets, by Proposition 1.2.4. Define $\tau : F \to H$ by setting $(\vee_I \wedge_J a_{ij}\pi)\tau = \vee_I \wedge_J a_{ij}\sigma$.

To show that τ is well-defined, suppose that $\vee_I \wedge_J a_{ij}\sigma \neq \vee_M \wedge_N b_{mn}\sigma$ are nonzero elements of $F\tau$. We shall now use repeatedly the distributive laws for the lattice F; we denote the set of functions from set Y to the set X by X^Y. Then,

$$\begin{aligned} 0 &\neq \vee_I \wedge_J a_{ij}\sigma \cdot \wedge_M \vee_N (b_{mn})^{-1}\sigma \\ &= \vee_I \wedge_J \vee_{N^M} \wedge_M (a_{ij}(b_{mr(m)})^{-1})\sigma \\ &= \bigvee_{I\cup(N^M)^J} \bigwedge_{J\cup M} (a_{ij}(b_{mq(r(m))})^{-1})\sigma. \end{aligned}$$

We have now expressed the latter element of the ℓ-group H in the normal form of Proposition 1.2.4. We shall now apply Lemma 6.2, which follows this proof. This lemma asserts that there exists a right ordering P_λ of G so that the corresponding word

$$\bigvee_{I\cup(N^M)^J} \bigwedge_{J\cup M} (a_{ij}(b_{mq(r(m))})^{-1})_\lambda$$

is nonzero, in the image of G in $A(G_\lambda)$ under the right regular representation. Therefore, by reversing the steps above, we have that $\vee_I \wedge_J (a_{ij})_\lambda \neq \vee_M \wedge_N (b_{mn})_\lambda$ in $A(G_\lambda)$. Consequently, $\vee_I \wedge_J a_{ij}\pi \neq \vee_M \wedge_N b_{mn}\pi$, and so τ is single-valued.

We shall now show that τ is a group homomorphism. Toward that end, suppose that $a = \vee_I \wedge_J a_{ij}\pi$ and $b = \vee_M \wedge_N b_{mn}\pi$ are elements of F. Then,

$$\begin{aligned}(ab^{-1})\tau &= \bigvee_{I\cup(N^M)^J} \bigwedge_{J\cup M} (a_{ij}(b_{mq(r(m))})^{-1})\sigma \\ &= \vee_I \wedge_J a_{ij}\sigma(\vee_M \wedge_N b_{mn}\sigma)^{-1} \\ &= a\tau(b\tau)^{-1}.\end{aligned}$$

Thus, τ is a group homomorphism.

Finally, we must show that τ is an ℓ-homomorphism. But for $a \in F$ expressed as above, we have

$$a \vee 1 = \vee_I \wedge_J (a_{ij}\pi \vee 1) = \vee_{I\cup\{z\}} \wedge_J a_{ij}\pi,$$

where $z \notin I$ and $a_{zj} = 1$ for all $j \in J$. Hence

$$\begin{aligned}(a \vee 1)\tau &= (\vee_{I\cup\{z\}} \wedge_J a_{ij}\pi)\tau \\ &= \vee_{I\cup\{z\}} \wedge_J a_{ij}\sigma \\ &= \vee_I \wedge_J (a_{ij}\sigma \vee 1) \\ &= (\vee_I \wedge_J a_{ij}\pi)\tau \vee 1 \\ &= a\tau \vee 1.\end{aligned}$$

Therefore, τ is an ℓ-homomorphism of F into H and $\pi \circ \tau = \sigma$. //

To complete the proof of Theorem 6.4 we need to prove the following lemma:

Lemma 6.5 Let G be a partially ordered group whose positive cone is an intersection of right orders, and let σ be an o-homomorphism from G into an ℓ-group H. Let $\vee_I \wedge_J a_{ij}\sigma \neq 1$, where I and J are finite index sets. Then, there exists a right order P_λ on G so that $1 \neq \vee_I \wedge_J (a_{ij})_\lambda$.

Proof: First, we shall extend the lattice order on H to a right order for which $\vee_I \wedge_J a_{ij}\sigma \neq 1$. We accomplish this as follows: Since H is an ℓ-group, we may as well assume that it is an ℓ-subgroup of $A(T)$, for some totally ordered set T. One of the following two cases holds:

Case (i): There exists $k \in I$ with $\wedge_J a_{kj}\sigma \not\leq 1$. Then there exists $t \in T$ with $t < t(\wedge_J a_{kj}\sigma)$. Now, well-order T so that t is the first element in the well-ordering; this determines a right order on $A(T)$ (and hence on H) which extends the given lattice order. Since in this ordering $\vee_J a_{kj}\sigma > 1$, it follows that $\vee_I \wedge_J a_{ij}\sigma > 1$.

Case (ii): $\wedge_J a_{ij}\sigma \leq 1$ for all $i \in I$. Then there exists $t \in T$ such that $t > t(\vee_I \wedge_J a_{ij}\sigma)$. Thus, for each $i \in I, t > t(\wedge_J a_{ij}\sigma)$. Choose a well-ordering for T for which t is the first element. This gives a right ordering of $A(T)$ for which $\wedge_J a_{ij}\sigma < 1$ for each $i \in I$. Thus, we have $\vee_I \wedge_J a_{ij}\sigma < 1$ in the right ordered group $A(T)$, and hence in H.

We shall now pull this right order on H back to one on G which extends its partial order, as follows. First, since $G/Ker(\sigma)$ is isomorphic to $G\sigma \subseteq H$, a right-ordered group, we can equip it with the induced right order. Then choose any right order P_λ on G extending its partial order, and consider the right order it induces on $Ker(\sigma)$. We can now define $g \in G$ to be positive if $g \notin Ker(\sigma)$ and Kg is positive in the right ordered group $G/Ker(\sigma)$, or, if g is positive in the right order we have chosen for $Ker(\sigma)$.

The right order we have chosen above is indeed one which extends the partial order on G; we call it P_λ. Furthermore, the natural map $G \to G/Ker(\sigma)$ is an o-homomorphism with respect to P_λ. Thus, $Ker(\sigma) \neq \vee_I \wedge_J (Ker(\sigma)a_{ij}) = Ker(\sigma) \vee_I \wedge_J a_{ij}$ since $\vee_I \wedge_J a_{ij} \neq 1$, and hence $\vee_I \wedge_J a_{ij} \neq 1$ in G_λ. Now embed G in $A(G_\lambda)$ via the right regular representation. By considering as two cases whether $\vee_I \wedge_J a_{ij}$ is positive or negative in G_λ, the reader can easily check that $\vee_I \wedge_J (a_{ij})_\lambda \neq 1$ in $A(G_\lambda)$. //

We can now apply our theory above to construct the free ℓ-group on a set of generators of arbitrary cardinality.

Corollary 6.6 There exists a unique free ℓ-group on a set of α generators, for any cardinal number α.

Proof: Let G be the free group on a set S of α generators, and equip G with the trivial partial order. Since G can be totally ordered (see E4), the positive cone $\{1\}$ of the trivially ordered G can be expressed as an intersection of right orders. Let F be the free ℓ-group on the partially ordered group G. We claim that F is the free ℓ-group on the set S of generators.

To show this, suppose that H is an ℓ-group, and μ is a map from S into H. Since G is free on S, we can lift μ to a map σ from G into H. Because G has the trivial order, σ is an o-homomorphism, and as such can be lifted to an ℓ-homomorphism from F into H, as required. //

Conrad's construction above of the free ℓ-group is a direct generalization of Weinberg's construction [63] of the free abelian ℓ-group on a partially ordered abelian group, whose positive cone is an intersection of total orders. In place of the $A(G_\lambda)$'s used above, Weinberg needed only copies of G, equipped with these various total orders. In particular, for a free abelian ℓ-group on a set of α generators, we have at each component a copy of the free abelian group on α generators, equipped with a total order.

Note of course that another way of describing the free abelian ℓ-group G on a set of generators is as an ℓ-homomorphic image of the free ℓ-group F on the same set of generators. Specifically, G is ℓ-isomorphic to F modulo the ℓ-ideal generated by the commutator subgroup of F. In fact, in this manner one can obtain a free ℓ-group for any variety of ℓ-groups.

It is interesting to note that the free ℓ-group on one generator is ℓ-isomorphic to a cardinal product of two copies of the additive integers. (This was first observed by Birkhoff [42].) For, if we place the trivial order on $\mathbf{Z}$, the free group on one generator, there are clearly only two right orders of $\mathbf{Z}$, where either $1 > 0$ or $-1 > 0$. (Of course, either of these right orders makes $\mathbf{Z}$ an o-group, since $\mathbf{Z}$ is abelian.) Let $\mathbf{Z}_1$ and $\mathbf{Z}_2$ be these two o-groups. The right regular representation of $\mathbf{Z}_i$ in $A(\mathbf{Z}_i)$ is ℓ-isomorphic to $\mathbf{Z}_i$. Hence, the free ℓ-group on one generator is ℓ-isomorphic to $\mathbf{Z} \boxplus \mathbf{Z}$, where $\mathbf{Z}$ has the usual order. The generator of this free ℓ-group is $(1, -1)$.

Since the free ℓ-group on one generator is therefore abelian, it is also the free abelian ℓ-group on one generator. We can thus think of this ℓ-group in terms of the Birkhoff-Baker-Beynon representation. We then obtain the set of real-valued functions fixing zero, which are linear to the left and to the right of zero, with integer slopes. The two slopes provide the obvious ℓ-isomorphism to $\mathbf{Z} \boxplus \mathbf{Z}$.

An alternative approach to the study of (nonabelian) free ℓ-groups is to use permutation groups. Glass [76a] proved that for regular infinite cardinals, the free ℓ-group F on that many generators admits an ℓ-embedding into the ℓ-group $A(T)$ of order-preserving permutations of the totally ordered set T, in such a way that F acts o-2 transitively on T, and for which no non-identity element of F has bounded support (F "acts pathologically" on T.) McCleary [85a],and independently Kopytov [83], extended Glass's result to all cardinals (greater than 1). Indeed, Kopytov's proof, as he observes, shows that the free ℓ-groupis orderable and so satisfies unique extraction of roots. Kopytov [83] has also shown that if a free ℓ-group admits a transitive representation on a totally ordered set, then there exists one of the right orders on the free group in the Conrad representation which gives a faithful representation of the free ℓ-group. The Glass-Kopytov-McCleary result then implies that this is the case for all free ℓ-groups on sets of cardinality greater than one. McCleary [85b] has since gone on to show that there exists a Kopytov right order on the free group upon which F acts both o-2 transitively and pathologically. Using the Glass-Kopytov-McCleary theorem, it is then easy to show that a free ℓ-group on more than one generator has many properties. For example, it has trivial center, is subdirectly irreducible, and is not completely distributive, since pathologically o-2 transitive ℓ-permutation groups possess these properties (see McCleary [73a]). These results are exceedingly difficult to obtain otherwise.

Chapter 7: *Varieties of ℓ-groups*

Section 7.1: *The lattice of varieties*

A variety for any type of abstract algebras is an equationally defined class. Such classes were first studied by Birkhoff [35], who proved in 1935 that a class of algebras is a variety exactly if it is closed under homomorphic images, subalgebras and direct products. B. H. Neumann [37] initiated their study for groups. This beginning led to considerable activity in the field of varieties of groups, with many results collected in H. Neumann's survey [N]. In the early 70's Jorge Martinez [72b],[74a] began the study of varieties of ℓ-groups, which has since become a very active area in the field. A particularly influential paper in the area is by Glass, Holland and McCleary [80].

In this chapter we will use multiplicative notation for the group operation, and will use the standard notations $[g,h]$ for the commutator $g^{-1}h^{-1}gh$, and g^h for the conjugate $h^{-1}gh$.

A *variety* of ℓ-groups is an equationally defined class, that is, the class of all ℓ-groups satisfying a collection of equations. This collection may contain an infinite number of equations. We will now describe more formally what it means for an ℓ-group to satisfy an equation. Let X be a countable set of symbols. An equation has the form $w = 1$ where w is a word of the form $\vee_I \wedge_J \Pi_K x_{ijk}$, where I, J and K are finite index sets, and each x_{ijk} is either an element x from the set of symbols, a formal inverse x^{-1} or an identity symbol 1. There are only finitely many symbols from X which occur in this word. A *substitution* for w in an ℓ-group G means the following: Replace all occurrences of each of the symbols which occur in the word by some element of G (and replace the identity by the identity for G). This gives us an element of G which we denote by $\overline{w}$. If $\overline{w} = 1$ in G for all possible substitutions in G, we say that G *satisfies* $w = 1$.

Of course, it would be rather cumbersome if we insisted on writing every equation in this formal manner. Complicated equations involving $\vee$ and $\wedge$ need some abbreviation in order to prevent the equations from being unwieldy. We can write $x \wedge y = x$ as $x \leq y$ and $x \vee x^{-1}$ as $|x|$. If in an equation we wish to have $1 \leq x$ we note this, instead of replacing x by $x \vee 1$ in the equation. More generally if we wish to have $1 \leq x \leq y$ we note this, instead of replacing x by $(x \wedge y) \vee 1$ and y by $x \vee y \vee 1$.

As an example of the shortening that results from the above, recall the inequality that defines the class of normal-valued ℓ-groups (Theorem 4.2.4): $xy \leq y^2x^2$ for all $1 \leq x$ and $1 \leq y$. Using basic properties of ℓ-groups (particularly the distributive laws), it is straightforward but tedious to put this in its formal equational form:

$$(1 \wedge x^2yx \wedge x^2y) \vee (1 \wedge xyx^{-1} \wedge y) \vee (1 \wedge yx^{-1} \wedge y) \vee$$
$$(1 \wedge xy) \vee (1 \wedge y) \vee (1 \wedge x^{-1}y) \vee (1 \wedge x^2yx^{-1} \wedge x^2) \vee$$
$$(1 \wedge xyx^{-1}) \vee (yx^{-1} \wedge 1) \vee (1 \wedge x \wedge x^2) \vee (1 \wedge x) \vee (x^{-1} \wedge 1) = 1.$$

(One can certainly obtain a shorter equation defining this variety if we do not insist that the equation be in formal form).

We will denote the variety of normal-valued ℓ-groups by $\mathcal{N}$. Other important varieties we have already met are these: the abelian variety $\mathcal{A}$ defined by the equation $xy = yx$; the representable variety $\mathcal{R}$ defined by $(x \wedge y)^2 = x^2 \wedge y^2$ (see Theorem 4.1.1); the variety $\mathcal{L}$ of all ℓ-groups defined by $x = x$; and the trivial variety $\mathcal{E}$ containing only the trivial one element ℓ-group, defined by $x = 1$. Results from Chapter 4 obviously imply that

$$\mathcal{L} \supset \mathcal{N} \supset \mathcal{R} \supset \mathcal{A} \supset \mathcal{E}$$

Suppose now that $\mathcal{V}$ is a variety defined by a set of equations $\{w_i = 1\}$. Let F be the free ℓ-group on countably many generators and I the ℓ-ideal of F generated by all possible substitutions of free generators into the words w_i. It is evident that $F/I \in \mathcal{V}$; in fact F/I is the free ℓ-group on countably many generators for the variety $\mathcal{V}$. (We have already seen this fact for the abelian variety in the discussion following Lemma 6.2.) Conversely suppose that I is an ℓ-ideal of F. For each $g \in I$, put g in the normal form of Proposition 1.2.4, in terms of the free generators of F. Then replace each free generator in this expression by a corresponding element of a symbol set X to obtain a word w. Then the set of all equations of the form $w = 1$ determines a variety whose free ℓ-group on countably many generators is F/I.

We thus have a one-to-one inclusion reversing correspondence between varieties of ℓ-groups, and ℓ-ideals of the free ℓ-group on countably many generators. This means that there are at most a continuum of varieties (we shall show there are exactly this many in the third section of this chapter). Furthermore, the set $\mathbf{V}$ of all varieties is a complete distributive lattice with the meet operation being simply intersection and the join operation defined as follows: Let U be a set of ℓ-group varieties; then

$$\bigvee U = \bigcap \{\mathcal{V} : \mathcal{V} \text{ is a variety and } \mathcal{V} \supseteq \mathcal{U} \text{ for all } \mathcal{U} \in U\}.$$

Birkhoff's theorem [35] characterizing varieties for any type of algebra is of vital importance for us; we shall record it here (expressed specifically as a theorem about ℓ-groups):

Theorem 7.1.1 Let $\mathcal{V}$ be a collection of ℓ-groups. Then $\mathcal{V}$ is a variety if and only if it has the following three properties:

(a) If $G \in \mathcal{V}$ and H is an ℓ-subgroup of G then $H \in \mathcal{V}$.

(b) If $G \in \mathcal{V}$, H is an ℓ-group and ϕ an ℓ-homomorphism with $G\phi = H$, then $H \in \mathcal{V}$.

(c) If $\mathcal{G}$ is a collection of ℓ-groups each of which is an element of $\mathcal{V}$, then $\prod \mathcal{G} \in \mathcal{V}$.

We sometimes express the content of this theorem by saying that a variety is a class of ℓ-groups closed with respect to ℓ-subgroups, ℓ-homomorphic images and direct products.

Suppose G is any non-trivial ℓ-group; then certainly the group of additive integers $\mathbf{Z}$ is ℓ-isomorphic to an ℓ-subgroup of G (consider any cyclic subgroup generated by a positive element of G). But a free abelian ℓ-group on any number of generators is a subdirect product of copies of $\mathbf{Z}$ (Theorem 6.1); any abelian ℓ-group is in turn an ℓ-homomorphic image of a free abelian ℓ-group. Consequently any variety having a nontrivial ℓ-group as an element must contain the abelian variety. That is, the abelian variety is the smallest proper variety of ℓ-groups (Weinberg explicitly observed this in his paper constructing free abelian ℓ-groups [63]); consequently it follows that any equation which $\mathbf{Z}$ does not satisfy is not satisfied by any nontrivial ℓ-group. It is interesting that the variety of distributive lattices plays a role for lattices analogous to $\mathcal{A}$; that is, it is the smallest proper variety of lattices.

It is remarkable to note that a "dual" statement holds; namely, the normal-valued variety $\mathcal{N}$ is the largest proper variety of ℓ-groups, proved by Charles Holland [76].

In order to prove this theorem we must discuss an important theorem of S. McCleary's on ordered permutation groups. We first require some definitions. Let H be an ℓ-subgroup of $A(T)$, the ℓ-group of order-preserving permutations of some totally ordered set T. Note that $A(T)$ can be naturally ℓ-embedded into $A(T^\wedge)$, where $T^\wedge$ is the Dedekind completion of T. An element $f \in A(T^\wedge)$ is a *period* for H if $fh = hf$ for all $h \in H$, and $\{tf^n\}$ is coterminal in T for all $t \in T$. We say that H is *periodic* on T if H acts transitively on T and H has a period.

We say that H is *o-primitive* if H acts transitively on T and H induces no non-trivial congruences on T (see [G] for definitions and details.) The latter is equivalent to the stabilizer subgroups $H_s = \{h \in H : sh = h\}$ being maximal convex ℓ-subgroups (See McCleary [72c] or [G].)

In the proof of Holland's theorem we first extract an o-primitive ℓ-group from a non-normal-valued ℓ-group. McCleary's theorem gives a description of o-primitive ℓ-groups which allows us then to deduce the existence of an o-2 transitive ℓ-group (Lemma 7.1.2), which is shown to satisify no non-trivial equation (Lemma 7.1.3). This proof is an example of the importance of McCleary's theorem in the study of o-permutation groups. Further important applications of McCleary's theorem can be found in the book by Glass [G].

McCleary's theorem states that an o-primitive ℓ-group H acting on T either acts o-2 transitively on T, is isomorphic to a right regular representation on a subgroup of $\mathbf{R}$ or is periodic on T, in which case H_s acts faithfully and o-2 transitively on the interval (s, sf) where f is a period of H. Note that in the second case $H_s = \{1\}$ for all $s \in T$. We therefore have the following:

Lemma 7.1.2 If the ℓ-subgroup H of $A(T)$ is o-primitive on T but $H_s \neq \{1\}$ for some $s \in T$, then H contains an ℓ-subgroup which is o-2 transitive on some totally set.

The following lemma is the other crucial ingredient to the proof of Holland's theorem since it asserts that an ℓ-group acting o-2 transitively on a totally ordered set cannot satisfy any nontrivial equation. In this technical but clever proof the action of a free ℓ-group (which of course satisfies no nontrivial equation) on a finite subset of a totally ordered set is mimicked in the o-2 transitive action of the given ℓ-group.

Lemma 7.1.3 An ℓ-group acting o-2 transitively on a totally ordered set satisfies no nontrivial ℓ-group equation.

Proof: Let S be a totally ordered set and suppose that H is an o-2 transitive ℓ-subgroup of $A(S)$. Let $w = 1$ be an equation not satisfied by some ℓ-group. Then this equation is not satisfied by F, the free ℓ-group on countably many generators. Hence there is a substitution for w by free generators of F which gives a nonidentity element. Let $u = \vee_I \wedge_J \Pi_K x_{ijk}$ be w with this substitution. We now assume that F is an ℓ-subgroup of $A(T)$ for some totally ordered set T. We know that $tu \neq t$ for some $t \in T$. We will mimic the action of u on T in the o-2 transitive action of H on S.

Suppose $K = \{1,2,\ldots,n\}$; then for each $(i,j) \in I \times J$, define $t(i,j,0) = t$ and for $1 \leq k \leq n$, $t(i,j,k) = t(i,j,k-1)x_{ijk}$. Now for each free generator x occurring in u and each (i,j), let $P_{ij}(x) = \{k : x = x_{ijk}\}$ and $N_{ij}(x) = \{k : x = (x_{ijk})^{-1}\}$. Then if $k \in P_{ij}(x)$ we have $t(i,j,k-1)x = t(i,j,k)$ and if $k \in N_{ij}(x)$ we have $t(i,j,k)x = t(i,j,k-1)$.

Now pick a subset of S that is order isomorphic to the finite subset $\{t(i,j,k)\}$ and label these points in an order-preserving fashion as $\{s(i,j,k)\}$. Since $t(i,j,k-1)x = t(i,j,k)$ for $k \in P_{ij}(x)$ and $t(i,j,k)x = t(i,j,k-1)$ for $k \in N_{ij}(x)$, we wish to find elements $h(x)$ of H so that $s(i,j,k-1)h(x) = s(i,j,k)$ for $k \in P_{ij}(x)$ and $s(i,j,k)h(x) = s(i,j,k-1)$ for $k \in N_{ij}(x)$. But H is o-2 transitive and hence o-n transitive (Theorem 5.8) and so such $h(x)$ exist. Since $t = t(i,j,0)$ for each (i,j), let $s = s(i,j,0)$. Then substituting $h(x)$ for x (and $h(x)^{-1}$ for x^{-1} and the identity of H for the identity of F) we have for each (i,j), $s\Pi_K h(x_{ijk}) = s(i,j,k)$. Since $tu \neq t$,

$$t \neq t \vee_I \wedge_J \Pi_K x_{ijk} = \vee_I \wedge_J t\Pi_K x_{ijk} = \vee_I \wedge_J t(i,j,k),$$

where in the last term we are taking the lattice operations on elements of the finite chain $\{t(i,j,k)\}$. But then, since $\{s(i,j,k)\}$ is order isomorphic to $\{t(i,j,k)\}$, we have

$$s \neq \vee_I \wedge_J s(i,j,k) = \vee_I \wedge_J s\Pi_K h(x_{ijk}) = s \vee_I \wedge_J \Pi_K h(x_{ijk}).$$

Consequently, the equation $w = 1$ is not satisfied in H, which completes the proof. //

We can now put these two Lemmas together to prove

Theorem 7.1.4 $\mathcal{N}$ is the largest proper ℓ-group variety.

Proof: Suppose that G is a non-normal-valued ℓ-group, and consider any equation not satisfied by every ℓ-group. If we can show that any variety containing G contains an ℓ-group that does not satisfy this equation, then the variety generated by G must satisfy only those equations satisfied by all ℓ-groups, in other words, the variety generated by G is $\mathcal{L}$ and hence $\mathcal{N}$ is the largest proper variety.

Now there exists a regular subgroup M of G such that M is not normal in its cover M^*. Let $K = \bigcap\{k^{-1}Mk : k \in M^*\}$. Then K is an ℓ-ideal of M^*; let $H = M^*/K$. Consider the right regular representation of this ℓ-group acting on the right cosets M^*/M. This is well-defined, since if $Kh = Kk$, $hk^{-1} \in K$, and so for any $m \in M^*$ we have $Mmhk^{-1} = Mm$, or $Mmh = Mmk$. We claim this representation is faithful: for if Kk is in the kernel of this map, then $Mm = Mmk$, for all $m \in M^*$, and so $k \in \bigcap m^{-1}Mm = K$. Consequently H is ℓ-isomorphic to an ℓ-subgroup of $A(M^*/M)$. In this representation the stabilizer of the coset $M1$ is M/K, and so we have $H_s \neq \{1\}$, for some s in the totally ordered set, since $K \subset M$. But then H_s is a maximal convex ℓ-subgroup of H and so H is o-primitive. Thus by Lemma 7.1.2 H contains an ℓ-subgroup which is o-2 transitive on some totally ordered set. Since this ℓ-subgroup is an ℓ-subgroup of an ℓ-homomorphic image of an ℓ-subgroup of G, it must be in any variety containing G. But by Lemma 7.1.3 such an ℓ-subgroup can belong to no proper variety. //

We can rephrase Holland's theorem as follows: Any ℓ-group satisfying an equation which is not satisfied by another ℓ-group is normal-valued; or equivalently, if any non-normal-valued ℓ-group satisfies an equation then every ℓ-group satisfies that equation.

We will now define multiplication of varieties. Let $\mathcal{U}$ and $\mathcal{V}$ be two varieties. The product $\mathcal{U}\mathcal{V}$ is the collection of all ℓ-groups G with ℓ-ideal H such that $H \in \mathcal{U}$ and $G/H \in \mathcal{V}$. (Notice that Martinez, who first discussed products of ℓ-group varieties, reversed this notation; we follow the notation of Neumann).

It is easy to show using Birkhoff's characterization of varieties that this product is a variety: For example, suppose that $\{G_\lambda\}$ is a collection of ℓ-groups belonging to the product variety; we shall denote by H_λ the ℓ-ideal of G_λ such that $H_\lambda \in \mathcal{U}$ and $G_\lambda/H_\lambda \in \mathcal{V}$. But then ΠH_λ is an l-ideal of ΠG_λ which belongs to $\mathcal{U}$ and $\Pi G_\lambda/\Pi H_\lambda$ is ℓ-isomorphic to $\Pi(G_\lambda/H_\lambda)$, which belongs to $\mathcal{V}$. The other two arguments are similar and left to the reader.

Notice that this multiplication of varieties is by no means commutative; however it follows that it is associative. Consequently, the set $\mathbf{V}$ of varieties is a semigroup.

It is easy to see that $\mathbf{V}$ is in fact a partially ordered semigroup; that is, if $\mathcal{U} \subseteq \mathcal{V}$ then

$$\mathcal{W}\mathcal{U} \subseteq \mathcal{W}\mathcal{V} \text{ and } \mathcal{U}\mathcal{W} \subseteq \mathcal{V}\mathcal{W}.$$

Furthermore each element of $\mathbf{V}$ is positive; that is, .

$$\mathcal{U}\mathcal{V} \supseteq \mathcal{V} \text{ and } \mathcal{V}\mathcal{U} \supseteq \mathcal{V}$$

for any varieties $\mathcal{U}$ and $\mathcal{V}$. It is in fact the case that $\mathbf{V}$ is a lattice-ordered semigroup, meaning that multiplication distributes from both sides over meet and join. We shall not prove these distributive laws here. Martinez [74a] showed that multiplication distributes from the right over finite meets and joins, and that it distributes from the left over meets [75b] for torsion classes; Smith [76] observed that this holds for varieties. Glass, Holland and McCleary [80] showed that multiplication distributes from the left over finite joins. In fact, all of these distributive laws hold for infinite meets and joins, except possibly for distributing from the left over infinite joins; this remains an open question (see Glass, Holland and McCleary [80]).

Note that because $\mathcal{N}$ is the largest proper variety, it follows that $\mathcal{N}^2$ is either $\mathcal{N}$ or $\mathcal{L}$. But $B(\mathbf{R})$, the ℓ-group of bounded order-preserving permutations of $\mathbf{R}$, is an ℓ-simple non-normal-valued ℓ-group (see E51), and consequently cannot belong to any product variety and hence not to $\mathcal{N}^2$. We thus have that $\mathcal{N}^2 = \mathcal{N}$; that is, $\mathcal{N}$ is *idempotent* in the semigroup **V**. We will later (Corollary 7.1.12) determine all idempotents in this semigroup.

Factoring in the semigroup **V** is more easily accomplished once we know that for a variety $\mathcal{V}$ and any ℓ-group G, G has a unique maximum ℓ-ideal (which is perhaps trivial) belonging to $\mathcal{V}$. We shall prove this in the following theorem, due to Holland [79]:

Theorem 7.1.5 Let $\mathcal{U}$ be a variety of ℓ-groups and G an ℓ-group. Then there exists a unique ℓ-ideal $\mathcal{U}(G)$ of G such that $\mathcal{U}(G) \in \mathcal{U}$ and $\mathcal{U}(G)$ contains every convex ℓ-subgroup of G belonging to $\mathcal{U}$.

Before proving the theorem we shall make two observations.

First, a very useful and immediate consequence of the theorem is that for varieties $\mathcal{U}$ and $\mathcal{V}$

$$G \in \mathcal{U}\mathcal{V} \text{ if and only if } G/\mathcal{U}(G) \in \mathcal{V}.$$

Second, note that a class of ℓ-groups closed under convex ℓ-subgroups, ℓ-homomorphic images and joins of convex ℓ-subgroups is called a *torsion class*; Theorem 7.1.5 can thus be rephrased as stating that every variety is a torsion class. We give some examples of torsion classes in the appendix.

To prove Theorem 7.1.5 we first need two Lemmas, the first of which is due to Martinez [75b] and the second to Holland [79]. Notice that the first of these lemmas is really just a special case of the theorem: namely, that the variety of normal-valued ℓ-groups is a torsion class.

Lemma 7.1.6 Suppose H is an ℓ-group and $\{G_\lambda\}$ is a set of normal-valued convex ℓ-subgroups of H. Let G be the convex ℓ-subgroup of H generated by $\{G_\lambda\}$. Then G is normal-valued.

Proof: Let N be a regular subgroup of G with cover N^*. Suppose that N is not normal in N^*. Then there exists $k \in N^* \backslash N$ with $N^k \neq N$. We may as well assume that k is positive. Using the fact that G is the group generated by the G_λ's (Theorem 1.2.2) and the Reisz decomposition property (Theorem 1.1.4) we know that k can be written as a product of positive elements from the G_λ's. Therefore there exists $h \in G_\lambda^+$ for some λ, for which $N^h \neq N$. Hence $G_\lambda \not\subseteq N$, and so by Theorem 1.2.13 $N \cap G_\lambda$ is a regular subgroup of G_λ covered by $N^* \cap G_\lambda$. But $h \in N^* \cap G_\lambda$, and so

$$N^h \cap G_\lambda = (N \cap G_\lambda)^h = N \cap G_\lambda,$$

since G_λ is normal-valued. But then by Theorem 1.2.13 $N = N^h$, which is a contradiction. Hence N is normal in N^*. //

Lemma 7.1.7 Let G be a subdirectly irreducible normal-valued ℓ-group generated by $g_1, \dots, g_n$. Then $G = G(g_j)$ for some $1 \leq j \leq n$.

Proof: Let C be a value of some element of the minimal ℓ-ideal of G. Then $\{g_1, \ldots, g_n\}$ is not contained in C. Let K be the largest member of the non-empty finite chain

$$\{M : C \subseteq M, M \text{ a value for some } g_i\}.$$

Then K is a value for g_j for some j and so K is normal in K^*. In fact, $K^* = G$ and so G/K is isomorphic to an ℓ-subgroup of the reals. Now let $h \in G^+\backslash K$. We wish to show that $h > k$, for all $k \in K$. It will then follow that $G = G(g_j)$.

So pick $k \in K^+$. Then $h \wedge k \leq k$ and so $k = \bar{k}(h \wedge k)$ where $\bar{k} \in K$ and $\bar{h} \wedge \bar{k} = 1$ where $\bar{h} = h(h \wedge k)^{-1}$. Now $\bigcap\{g^{-1}Cg : g \in G\}$ is an ℓ-ideal of G which clearly does not contain the minimal ℓ-ideal of G, and so $\bigcap\{g^{-1}Cg : g \in G\} = \{1\}$. Since C is a value each of its conjugates is prime. If $\bar{h} = h(h \wedge k)^{-1} \in g^{-1}Cg$ then $h(h \wedge k)^{-1} \in K$ implying that $h \in K$ which is a contradiction, so $\bar{h} \notin g^{-1}Cg$. Hence $\bar{k} \in \bigcap\{g^{-1}Cg : g \in G\} = \{1\}$, and therefore $h \wedge k = k$. //

Proof of Theorem 7.1.5: Let H be an ℓ-group and $\{G_\lambda\}$ a collection of convex ℓ-subgroups of H such that each member satisfies the equation $w = 1$. If every ℓ-group satisfies $w = 1$ then certainly so does $\vee G_\lambda$. If not then every G_λ must be normal-valued. Now suppose the ℓ-group word w involves the symbols $x_1, \ldots, x_m$. Let $h_1, \ldots, h_m \in \vee G_\lambda$. Let $\bar{w}$ be the substitution for w obtained by replacing each x_i by h_i. We wish to show that $\bar{w} = 1$. Since $\vee G_\lambda$ is just the subgroup generated by the G_λ's, $h_i = \Pi g_{ij}$ for $\{g_{ij}\} \subseteq \cup G_\lambda$. Let G be the ℓ-subgroup of $\vee G_\lambda$ generated by $\{g_{ij}\}$. Since G is an ℓ-subgroup of a normal-valued ℓ-group it is also normal-valued. Let $\bar{G}$ be any subdirectly irreducible factor of G and denote the natural map from G to $\bar{G}$ by $g \mapsto \bar{g}$. Then $\bar{G}$ is normal-valued and generated by $\{\bar{g}_{ij}\}$. Therefore by Lemma 7.1.7 $\bar{G} = \bar{G}(\bar{g}_{kl})$ for some k and l. Since $g_{kl} \in G_\lambda$ for some λ, and since $\overline{G_\lambda \cap G}$ is a convex ℓ-subgroup of $\bar{G}$, $\bar{G} = \overline{G_\lambda \cap G}$. Because G_λ satisfies $w = 1$, so does $\overline{G_\lambda \cap G} = \bar{G}$. Finally, G is a subdirect product of subdirectly irreducible factors, each of which satisfies $w = 1$; hence G does also. In particular $\bar{w} = 1$.//

We shall now consider the varieties obtained as powers of the abelian variety. For example, $\mathcal{A}^2$ is the variety of ℓ-groups that have an abelian ℓ-ideal with abelian quotient (*metabelian* ℓ-groups). Clearly $\mathcal{A}^n$ is properly contained in $\mathcal{A}^{n+1}$ (we will give explicit examples below), and since $\mathcal{N}$ is the largest proper variety $\mathcal{A}^n \subseteq \mathcal{N}$ for all positive integers n. Thus we have

$$\mathcal{A} \subset \mathcal{A}^2 \subset \mathcal{A}^3 \subset \ldots \subset \mathcal{N}.$$

Hence $\bigvee \mathcal{A}^n$ is contained in $\mathcal{N}$; an interesting and important result due to Glass, Holland and McCleary [80] is that $\bigvee \mathcal{A}^n = \mathcal{N}$. The proof of this theorem is our next objective.

In order to prove this we need to introduce the concept of wreath product, an important construction for many other reasons. The reader should refer to Holland and McCleary [69] or [G] for a more detailed account. Neumann[60] first constructed wreath products of o-groups. Lloyd [64] and Reilly [69] were the first to consider the wreath product of two ℓ-groups; Holland and McCleary extended this to any cardinality. We shall begin by describing the wreath product of two ℓ-groups, and then generalize that to the case of an arbitrary number of o-groups. This can be extended to an arbitrary number of ℓ-groups, but we shall not need that full generality here; the reader should consult Holland and McCleary [69] for details.

Let G and H be ℓ-groups. We shall need a faithful representation of G as a group of permutations of a totally ordered set T; we may as well assume that G is actually an ℓ-subgroup of $A(T)$. We now let

$$W = \{(\langle h_t \rangle, g) : g \in G, h_t \in H \text{ for all } t \in T\},$$

where we think of $\langle h_t \rangle$ as a vector with components from H indexed by T. We now define the customary binary operation on W by $(\langle h_t \rangle, g)(\langle k_s \rangle, d) = (\langle a_t \rangle, gd)$, where $a_t = h_t k_{tg}$, for all $t \in T$. It is easy to show that this makes W a group with identity $(\langle 1_t \rangle, 1)$, where 1_t is the identity of H for all $t \in T$, and $(\langle h_t \rangle, g)^{-1} = (\langle k_t \rangle, g^{-1})$, where $k_t = (h_{tg^{-1}})^{-1}$, for all $t \in T$. We can partially order W by saying that $(\langle h_t \rangle, g)$ is positive if and only if $g \geq 1$ and $h_t \geq 1$ for all $t \in T$ for which $tg = t$. One can easily show that W with this order is an ℓ-group. We call W the *(large) wreath product* of H by G (acting on T), which we write as H Wr $G(T)$; we shall often suppress mention of the totally ordered set T and write H Wr G instead.

Now consider those elements $(\langle h_t \rangle, g)$ of H Wr G for which $h_t \neq 1$ for only finitely many t. The set of these elements is an ℓ-subgroup of the (large) wreath product called the *small wreath product* and written as H wr G.

The set of elements $(\langle h_t \rangle, g)$ of H Wr G (or the small wreath product) where $g = 1$ is called the *natural ℓ-ideal* of the wreath product. This is easily shown to be ℓ-isomorphic to the cardinal product (sum) of $|T|$ copies of H, and is consequently denoted ΠH (ΣH). Notice that the wreath product modulo the natural ℓ-ideal is ℓ-isomorphic to G. From this observation it follows easily that if H belongs to variety $\mathcal{U}$ and G belongs to variety $\mathcal{V}$ then H Wr G belongs to variety $\mathcal{U}\mathcal{V}$.

An important special case of this construction occurs when G is an o-group; in fact, this is the only case we will be using in this chapter. We can then use the usual right regular representation of G acting on itself; that is, we can let T be G. We then have that

$$W = \{(\langle h_f \rangle, g) : g \in G, h_f \in H \text{ for all } f \in G\}.$$

It is important to note that even if both G and H are o-groups the resulting wreath product is an ℓ-group and not an o-group. In fact it is impossible to totally order W (see E1).

We will now generalize the construction above, by obtaining the wreath product of an arbitrary collection of o-groups, indexed by some totally ordered set. Holland and McCleary [69] actually further generalize to obtain a wreath product of an arbitrary collection of ℓ-groups, by viewing each such ℓ-group as a group of permutations of a chain; we will not need this added complication here.

Let Γ be a totally ordered index set, and for each $\gamma \in \Gamma$ let G_γ be an o-group whose identity element is 1_γ. Let

$$V = \{r \in \prod G_\gamma : r \text{ has inversely well-ordered support }\}.$$

Order V lexicographically from above: for $r, s \in V$, $r < s$ if and only if $r_{\gamma_0} < s_{\gamma_0}$ in G_{γ_0} where γ_0 is the largest element in $\{\gamma \in \Gamma : r_\gamma \neq s_\gamma\}$. This makes V a totally ordered set (and in fact is just the Hahn group $V(\Gamma, G_\gamma)$ indexed by Γ with the γth entry from G_γ).

For each $\gamma \in \Gamma$, define an equivalence relation K^γ on V by setting $rK^\gamma s$ if and only if $r_\alpha = s_\alpha$ for all $\alpha > \gamma$. Note that if $\gamma > \delta$ then $K^\gamma \supseteq K^\delta$.

We can now define the *wreath product* $W = \mathrm{Wr}\{G_\gamma : \gamma \in \Gamma\}$ as consisting of all $\Gamma \times V$ "matrices" $(g_{\gamma,r})$ with entries $g_{\gamma,r} \in G_\gamma$ whose "rows" satisfy the condition $g_{\gamma,r} = g_{\gamma,s}$ if $rK^\gamma s$, and whose "columns" have inversely well-ordered support (that is, the "columns" are elements of V).

It is easiest to understand the group operation of W by viewing it as an ℓ-permutation group acting on the totally ordered set V. Namely, we consider the following action of W on V:

$$(s(g_{\gamma,r}))_\delta = s_\delta g_{\delta,s},$$

where s_δ is the δth component of s. In other words, an element w of W operates on $s \in V$ by multiplying it componentwise by the sth "column" of w. Note that this would not necessarily give another element of V were the restriction given above on the "columns" of W omitted.

This action certainly makes W a subset of $A(V)$; and in fact the reader can check that W is an ℓ-subgroup with group operation given by $(g_{\gamma,r})(f_{\gamma,r}) = (h_{\gamma,r})$ where $h_{\gamma,r} = g_{\gamma,r} f_{\gamma, r g_{\gamma,r}}$.

We can now define the *small wreath product* $\mathrm{wr}\{G_\gamma : \gamma \in \Gamma\}$ as the ℓ-subgroup of W consisting of those "matrices" $(g_{\gamma,r})$ in which all entries are 1_γ except for finitely many γ, and for each γ, $g_{\gamma,r} = 1_\gamma$ except on at most finitely many K^γ-classes.

Note that if Γ is a two element chain we obtain the wreath product of o-groups we originally described. If Γ has n elements and each G_γ equals G, we write the wreath product as $\mathrm{Wr}^n G$ and the corresponding small wreath product as $\mathrm{wr}^n G$. If Γ is the set of positive integers we will write $\mathrm{Wr}^\infty G$ and $\mathrm{wr}^\infty G$ respectively. Using this notation we observe the following:

Lemma 7.1.8 $\mathrm{Wr}^n \mathbf{Z}, \mathrm{Wr}^n \mathbf{R} \in \mathcal{A}^n \backslash \mathcal{A}^{n-1}$, for $n > 1$.

Proof: We have observed that if $H \in \mathcal{U}$ and $G \in \mathcal{V}$ then $H \mathrm{\ Wr\ } G \in \mathcal{UV}$. Clearly neither $\mathrm{Wr}^2\mathbf{Z}$ nor $\mathrm{Wr}^2\mathbf{R}$ is abelian, and so $\mathrm{Wr}^2\mathbf{Z}, \mathrm{Wr}^2\mathbf{R} \in \mathcal{A}^2 \backslash \mathcal{A}$. Induction yields the lemma. //

Holland and McCleary were able to show that every o-group can be ℓ-embedded into a wreath product of real numbers, thus generalizing Hahn's representation theorem for abelian o-groups (see Chapter 3). (B.H. Neumann did this for solvable o-groups.) This theorem is not entirely satisfactory, of course, because the wreath product of a set of totally ordered groups is no longer totally ordered but only lattice-ordered. Read [74] has extended this result to normal-valued ℓ-groups:

Theorem 7.1.9 Every normal-valued ℓ-group can be ℓ-embedded into a wreath product $\mathrm{Wr}\{\mathbf{R}_\gamma : \gamma \in \Gamma\}$, where each $\mathbf{R}_\gamma$ is the group of real numbers.

We will not prove this theorem here (see [G] Theorems 5A, 11A and 11B).

We are now ready to prove the following theorem:

Theorem 7.1.10 $\mathcal{N} = \bigvee\{\mathcal{A}^n : n = 1, 2, \ldots\}$.

Proof: We have already observed that $\bigvee \mathcal{A}^n \subseteq \mathcal{N}$. Now suppose that $w = 1$ is an equation which fails to hold for some normal-valued ℓ-group and hence for some subdirectly irreducible normal-valued ℓ-group. We need to show that $w = 1$ fails in some abelian power $\mathcal{A}^n$. By Read's Theorem 7.1.9, the equation $w = 1$ fails in a wreath product $W = \mathrm{Wr}\{\mathbf{R}_\gamma : \gamma \in \Gamma\}$. We wish to show that $w = 1$ in fact fails in a wreath product of reals over a finite index set. By Lemma 7.1.8 this wreath product will be an element of one of the abelian powers.

Since $w = 1$ fails in W, which can be considered an ℓ-subgroup of $A(T)$ for a totally ordered set T, there is a substitution of elements in W for the symbols in the word w and a point $r \in T$ such that $r(\vee_I \wedge_J \Pi_K g_{ijk}) \neq r$. We shall now use a technique similar to that used to prove that $\mathcal{N}$ is the largest proper variety (Theorem 7.1.3). In particular, we consider the finite set $S \subset R$ consisting of r together with all points

$$rg_{ij1}, rg_{ij1}g_{ij2}, \ldots, rg_{ij1}\cdots g_{ijk},$$

for each $i \in I, j \in J, k \in K$. (We can think of S as all the points r "visits" as the permutations $\Pi_K g_{ijk}$ are applied.) For each pair s_1, s_2 of distinct points of S, there is a unique γ such that s_1 and s_2 differ at γ but $s_1 K^\gamma s_2$. Let Δ be the finite set of all such γ. Let $T' = \Pi\{\mathbf{R}_\delta : \delta \in \Delta\}$ and let $s \mapsto s'$ be the projection of T onto T'. This projection is one-to-one on S. We will now define a collection of elements of the finite wreath product $\mathrm{Wr}\{\mathbf{R}_\delta : \delta \in \Delta\}$ whose substitution into the word w will move s'.

For each $g \in \{g_{ijk}\}$ define $g' = (g'_{\gamma,t})$ where $g'_{\gamma,t}$ is $g_{\gamma,r}$ if there exists $p \in S$ such that $rgK^\gamma p$ in T', and 1 otherwise. One can verify that each g' is a well-defined element of the finite wreath product. Furthermore if we replace g_{ijk} by $(g_{ijk})'$ in the word w, we obtain an element of the finite wreath product which moves s'. //

Corollary 7.1.11 If $\mathcal{V}$ is a proper variety of ℓ-groups then $\bigvee \mathcal{V}^n = \mathcal{N}$.

Proof: Since $\mathcal{A} \subseteq \mathcal{V} \subseteq \mathcal{N}$ and $\mathcal{N}^2 = \mathcal{N}$ it follows that $\mathcal{A}^n \subseteq \mathcal{V}^n \subseteq \mathcal{N}$ for all positive integers n. Hence $\bigvee \mathcal{A}^n = \bigvee \mathcal{V}^n = \mathcal{N}$. //

Corollary 7.1.12 The only idempotent varieties are $\mathcal{E}$, $\mathcal{N}$ and $\mathcal{L}$.

The above discussion on the abelian powers allows for the concept of *dimension* of a variety. Namely, $\dim \mathcal{V} = n$ if $\mathcal{V} \supseteq \mathcal{A}^n$ but $\mathcal{V} \not\supseteq \mathcal{A}^{n+1}$.

Section 7.2: *Covers of the abelian variety*

The abelian variety $\mathcal{A}$ is defined by the single equation $xy = yx$. Every variety that is defined by a single equation is not the intersection of a descending tower of varieties properly containing it. One proof of this uses the Compactness Theorem of mathematical logic which says that if we have a set of first order sentences such that every finite subset of this set is satisfiable (that is, there is a model which satisfies all the sentences in this subset), then the entire set is satisfiable. (See for example [Ba].)

For suppose that $\mathcal{V}_1 \supset \mathcal{V}_2 \supset \ldots$ is a descending tower of varieties (of any type of algebra). Assume that $\mathcal{V} = \cap \mathcal{V}_i$, is defined by a single equation e. Now for each i, there must be an algebra in $\mathcal{V}_i$ that does not satisfy e. That is, there is an algebra which satisfies all of the equations that define $\mathcal{V}_i$ and, in addition, satisfies $\neg e$ (the negation of e). It follows from the compactness theorem that there is an algebra that satisfies all of the equations that any of the $\mathcal{V}_i$ satisfy and also $\neg e$, which is a contradiction.

Therefore, it follows that $\mathcal{A}$ has covers in the lattice of ℓ-group varieties. (Using more sophisticated methods, Gurchenkov [84] has shown that all proper ℓ-group varieties have covers.) In this section we will examine some of these covers.

We start with the varieties of Scrimger [75]. This is a countably infinite collection of covers of $\mathcal{A}$, none of which contain any nonabelian representable ℓ-groups.

For each positive integer n, we consider the subsets G_n of $\mathbf{Z}$ Wr $\mathbf{Z}$ defined by

$$G_n = \{(\langle w_i \rangle, \bar{w}) : i \equiv j \pmod{n} \text{ implies } w_i = w_j\}.$$

If $u = (\langle u_i \rangle, \bar{u})$ and $v = (\langle v_i \rangle, \bar{v})$ are elements of G_n, and if $i \equiv j \pmod{n}$, it is straightforward to show that $(uv)_i = (uv)_j$, $(u^{-1})_i = (u^{-1})_j$ and $(u \vee 1)_i = (u \vee 1)_j$. Thus, G_n is an ℓ-subgroup of $\mathbf{Z}$ Wr $\mathbf{Z}$.

For each positive integer n, let $\mathcal{L}_n$ denote the variety of ℓ-groups satisfying the equation $x^n y^n = y^n x^n$ (these varieties were first studied by Martinez [72b]). One can easily show that $G_n \in \mathcal{L}_n$.

Now let a and x denote the elements of G_n given by:

$$\bar{a} = 0 \quad , \quad a_i = \begin{cases} 1 & \text{if } i \equiv 0 \pmod{n} \\ 0 & \text{if } i \not\equiv 0 \pmod{n}. \end{cases}$$

$$\bar{x} = 1 \quad , \quad x_i = 0, \text{ forall } i.$$

Then, $\overline{a^m x^m} = m = \overline{x^m a^m}$, and

$$(a^m x^m)_i = \begin{cases} m & \text{if } i \equiv 0 \pmod{n} \\ 0 & \text{if } i \not\equiv 0 \pmod{n}. \end{cases}$$

$$(x^m a^m)_i = x^{m_i} + a^{m_{z+m}} = a^{m_{z+m}} = \begin{cases} m & \text{if } z + m \equiv 0 \pmod{n} \\ 0 & \text{if } z + m \not\equiv 0 \pmod{n}. \end{cases}$$

Therefore, $x^m a^m = a^m x^m$ if and only if $m \equiv 0 \pmod{n}$, that is, if and only if n divides m. Consequently, if n is a proper divisor of m, then m does not divide n and so G_m is not an element of $\mathcal{L}_n$. In addition, if $\mathcal{L}_n \subseteq \mathcal{L}_m$, then $G_n \in \mathcal{L}_m$, and so $a^m x^m = x^m a^m$, which implies that m divides n.

Let $\mathcal{S}_n$ be the variety generated by G_n. We will show that if n is prime, then $\mathcal{S}_n$ covers $\mathcal{A}$. This will be done by showing that every nonabelian subdirectly irreducible ℓ-group in $\mathcal{L}_n$ contains an ℓ-subgroup ℓ-isomorphic to G_n.

Lemma 7.2.1 If C is a convex ℓ-subgroup of $G \in \mathcal{L}_n$, then $x^{-n}Cx^n = C$, for all $x \in G$.

Proof: Suppose $1 < c \in C$. Then $1 < c < c^2 < \ldots < c^n = x^{-n}c^n x^n \in x^{-n}Cx^n$, and so $c \in x^{-n}Cx^n$. Hence, $C \subseteq x^{-n}Cx^n$, and similarly $C \subseteq x^n C x^{-n}$. Thus, $C = x^{-n}Cx^n$. //

Lemma 7.2.2 If C is a convex ℓ-subgroup of $G \in \mathcal{L}_n$ and $x \in G$, then the number of distinct conjugates of C of the form $x^{-i}Cx^i$ is a divisor of n.

Proof: Let i be the smallest positive integer such that $x^{-i}Cx^i = C$. If i is not a divisor of n there are integers r and s such that $rn + si = k$, where $1 \leq k < i$ and k is the greatest common divisor of n and i. Then $x^{-k}Cx^k = x^{-rn-si}Cx^{rn+si} = C$, contradicting the minimality of i. //

Lemma 7.2.3 (Martinez [74a]) For any positive integer n, $\mathcal{R} \cap \mathcal{L}_n = \mathcal{A}$.

Proof: Let G be a subdirectly irreducible ℓ-group in $\mathcal{R} \cap \mathcal{L}_n$. Then G is totally ordered. Suppose G is nonabelian. Then there exists a and b in G such that $b^{-1}ab < a$. Hence, $b^{-1}a^n b < a^n$ and so

$$a^n = b^{-n}a^n b^n < b^{1-n}a^n b^{n-1} < \ldots < b^{-1}ab < a^n,$$

which is a contradiction. Therefore G is abelian and so $\mathcal{R} \cap \mathcal{L}_n = \mathcal{A}$. //

Lemma 7.2.4 If C is a representing subgroup of $G \in \mathcal{L}_n \backslash \mathcal{A}$ then there exists $1 < x \in G$ such that $x^{-1}Cx \neq C$.

Proof: Suppose $x^{-1}Cx = C$ for all $1 < x \in G$. Then C is an ℓ-ideal of G. Since C is a representing subgroup, it follows that C is trivial, and so G is totally ordered and therefore representable. But $\mathcal{R} \cap \mathcal{L}_n = \mathcal{A}$, which gives a contradiction. //

The following lemma is due to Smith [80]. While the result is not very surprising, the proof is tedious and is omitted:

Lemma 7.2.5 Suppose a and x are positive elements of an ℓ-group G such that $1 < a < x$, $x^{-i}ax^i \wedge a = 1$ for $i = 1, 2, \ldots, n-1$, and $x^{-n}ax^n = a$. Then G contains an ℓ-isomorphic copy of G_n as an ℓ-subgroup.

The following is the key lemma for showing that the varieties $\mathcal{S}_n$ cover $\mathcal{A}$.

Lemma 7.2.6 If C is a representing subgroup of $G \in \mathcal{L}_n$ which has n distinct conjugates of the form $x^{-i}Cx^i$ for some $1 < x \in G$, then G contains an ℓ-subgroup ℓ-isomorphic to G_n.

Proof: Suppose we have G,C and x with the above properties. We will show that there exist two elements of G with the properties given in Lemma 7.2.5. For $0 \leq i \leq n-1$, define $C_i = x^{-i}Cx^i$ and $D_i = \bigcap\{C_j : j \neq i\}$.

Let S be the chain of right cosets of $C_0 = C$ and let s_0 be the coset C_0. For $0 < i \leq n-1$ define $s_i = s_0x^i = Cx^i$. It follows that C_i is the stabilizer of s_i, when translating S by right multiplication, and so D_i consists of all permutations of S which fix $\{s_j : j \neq i\}$. By hypothesis, $C_0 \neq C_i$ for $i = 1, 2, \ldots, n-1$; therefore there exist $h_i \in C_0^+ \backslash C_i$ for $i = 1, \ldots, n-1$. Hence

$$g = h_1h_2 \ldots h_{n-1} \in C_0 \backslash \bigcup_{i=1}^{n-1} C_i.$$

Define $d_0 = x \wedge \left(\bigwedge_{i=1}^{n-1} x^{-i}g^nx^i\right)$. Now, g fixes only s_0 among $s_0, \ldots, s_{n-1}$; hence $x^{-i}g^nx^i$ fixes only s_i among the s_j's . Therefore, $\bigwedge_{i=1}^{n-1} x^{-i}g^nx^i$ moves only s_0 among the s_j's . Also, x moves s_0 and so d_0 moves only s_0 among the s_j's . Therefore, $d_0 \in D_0 \backslash \bigcup_{i=1}^{n-1} D_i$. So, if we define $d_i = x^{-i}d_0x^i$ for $i = 1, \ldots, n-1$, we have that $d_i \in D_i \backslash \bigcup\{D_j : j \neq i\}$.

Define $e = \bigwedge_{j \neq 0}(d_0 \wedge d_j)$. Now, $e \in \bigcap_{j=0}^{n-1} D_j = \bigcap_{j=0}^{n-1} C_j$, by convexity. Furthermore, $1 \leq e < d_0$, since $d_0 \notin \bigcup\{D_j : j \neq i\}$. Let $a = d_0e^{-1}$. Then, straightforward calculations verify that $x^{-n}ax^n = a$, $1 < a < x$, and $x^{-i}ax^i \wedge x^{-j}ax^j = 1$ if $i \neq j$ and $i, j < n$. The lemma then follows from Lemma 7.2.5. //

Other facts about the varieties $\mathcal{L}_n$, S_n and some related varieties can be found in Scrimger [75], Smith [81], Gurchenkov [84b] and Holland and Reilly [86].

Notice that the varieties S_n, for n prime, give a countable collection of covers of $\mathcal{A}$. However, none of these covers are contained in the representable variety. In fact, Darnel [86] and Reilly [86] have independently proved that the Scrimger varieties are the only nonrepresentable covers of $\mathcal{A}$. (Kopytov also has announced this result.) We will prove this theorem in Chapter 10, after we have developed the needed machinery.

We now turn to representable covers of the abelian variety; we will first describe the Medvedev varieties, three covers of $\mathcal{A}$ generated by o-groups and consequently contained in the representable variety.

For that purpose, let

$$N = \langle a, b, c : c = [a, b], [a, c] = 1 = [b, c] \rangle,$$

be the free two generator nilpotent class two group. Notice that any element of N can be written uniquely in the form $a^mb^nc^k$. We can then totally order N lexicographically as follows: $a^mb^nc^k > 1$ if and only if $m > 0$, or $m = 0$ and $n > 0$, or $m = n = 0$ and $k > 0$. Thus, $a \gg b \gg c > 1$. We denote N equipped with this total order by N_0.

Theorem 7.2.7 Every variety containing a non-abelian nilpotent ℓ-group contains N_0. Consequently, the variety generated by N_0 covers $\mathcal{A}$.

Proof: Let $\mathcal{V}$ be a variety containing a non-abelian nilpotent ℓ-group F. The variety generated by F consists of nilpotent ℓ-groups and is contained in the representable variety $\mathcal{R}$ (see Theorem 4.1.2). Hence, its subdirectly irreducible elements are o-groups. Therefore, $\mathcal{V}$ must contain a non-abelian nilpotent class 2 o-group G on two generators, say a and b. Let $c = [a,b]$. Then c is in the center of of G and $c \neq 1$. Now, suppose that G satisfies some non-trivial group equation. Any substitution of elements of G into this group equation reduces to an equation of the form $a^m b^n c^k = 1$. Hence, $a^m b^n = c^{-k} = b^n b^{-n} c^{-k} = b^n c^{-k} b^{-n} = b^n a^m$, and so $[a^m, b^n] = 1$. But since c is in the center of G we have that $[a^m, b] = c^m$ and $[a, b^n] = c^n$, and so $[a^m, b^n] = c^{mn} = 1$. But G is an o-group, and so $c = 1$, which is a contradiction. It follows that G is N with some total order.

Without loss of generality, we may assume that $a, b > 1$ and $c > 1$ (if $c < 1$, then replace c by $[b,a]$). Since $[ba, b] = [a,b]$, we may also assume that $a > b$. Then, $c < a^{-1}b^{-1}ab < b$. Suppose that $b \leq c^r$, where r is the smallest such positive integer. Then $bc = aba^{-1} \leq ac^r a^{-1} = c^r$; so $b \leq c^{r-1}$, which is a contradiction. Thus, $1 < c \ll b$. Now consider $(1, a, a^2, \ldots)$ and $(b, b, b, \ldots)$, elements of ΠG. Let a' and b' be the respective images of the above two elements when ΣG is factored out of ΠG. Then we have that $a' \gg b' \gg [a', b'] > 1$ in the factor group H. Since N is free nilpotent of class two on a and b, the map $a \mapsto a'$, $b \mapsto b'$ extends to a homomorphism $\phi : N \to H$ which is clearly an ℓ-homomorphism from N_0 onto the ℓ-subgroup K of H generated by a' and b'. But the above argument shows that K is free nilpotent of class two, and so N_0 is ℓ-isomorphic to K. Since $H \in \mathcal{V}$ and K is an ℓ-subgroup of $H, K \in \mathcal{V}$ and so $N_0 \in \mathcal{V}$. //

To obtain the other two Medvedev varieties, we equip the group $\mathbf{Z}$ wr $\mathbf{Z}$ with two total orders, as follows: If $g = (\langle g_i \rangle, \bar{g}) \in \mathbf{Z}$ wr $\mathbf{Z}$, we call g positive if $g > 0$, or if $g = 0$ and $g_i > 0$ where i is the maximum element of the support of $\langle g_i \rangle$. Denote $\mathbf{Z}$ wr $\mathbf{Z}$ equipped with this order by M^+. Alternatively, we could alter the definition above by requiring that i be the minimum element of the support of $\langle g_i \rangle$. We denote $\mathbf{Z}$ wr $\mathbf{Z}$ equipped with this order by M^-.

Denote by b the element of $\mathbf{Z}$ wr $\mathbf{Z}$ where $\bar{b} = 0$, $b_0 = 1$ and $b_i = 0$, for all $i \neq 0$; denote by a the element for which $\bar{a} = 1$ and $a_i = 0$, for all i. Notice that in M^+, $1 < b \ll b^a \ll a$ and in M^-, $1 < b^a \ll b \ll a$.

We are now ready to show that M^+ and M^- generate representable covers of $\mathcal{A}$:

Theorem 7.2.8 The varieties generated by M^+ and M^- are covers of $\mathcal{A}$ which are contained in $\mathcal{R}$.

Proof: We shall prove this theorem for M^+; the argument for M^- is similar. The reader may observe that these varieties are distinct, and contained in the representable variety (since M^+ and M^- are o-groups).

Suppose that $\mathcal{V}$ is a variety properly containing $\mathcal{A}$, and contained in the variety generated by M^+. We can then choose a non-abelian ℓ-group $G \in \mathcal{V}$; from Birkhoff's theorem, there is an index set Λ such that $\Pi_\Lambda M^+$ contains an ℓ-subgroup H, and an onto ℓ-homomorphism $\phi : H \to G$. Let g_1, g_2 be non-commuting elements of G and $h_i \in H$ such that $h_i\phi = g_i$, for $i = 1,2$. Let $u = |[h_1, h_2]|$ and $v = |h_1||h_2|$.

If for $\lambda \in \Lambda$, v_λ is such that $\bar{v}_\lambda = 0$, then $(\bar{h}_1) = (\bar{h}_2) = 0$ and in this case $u_\lambda = 0$. On the other hand, $\bar{u}_\lambda = 0$. Thus, in either case we have that $u_\lambda \ll v_\lambda$ (and so $u \ll v$). In a similar fashion, $u \ll u^v$.

Now, let K be the ℓ-subgroup of H generated by u and v, and consider the map ψ from M^+ to K that takes b to u and a to v; ψ is an ℓ-isomorphism. Furthermore, if $0 < g \in K$ then there exists $n \in \mathbf{Z}$ such that $u^{v^n} < g$. But this means that ϕ cannot send g to the identity, and consequently $\phi|_K$ is one-to-one. Thus, we have $M^+ \cong K \cong K\phi$, which is an ℓ-subgroup of G. Hence, $M^+ \in \mathcal{V}$. //

In addition, Medvedev [83b] has shown that any variety which contains a solvable representable non-abelian ℓ-group must contain one of the three varieties described above.

There are also other covers of $\mathcal{A}$ contained in $\mathcal{R}$. Consider the following inequality (which could be put in equational form):

$$(*) \qquad |[a,b]| \le \big|\big[|[a,b]|, |[a,b]|^a\big]\big|, \text{ for } 1 \le b \le a.$$

Notice that (*) is trivially satisfied by any abelian ℓ-group. It is also easy to show that the three Medvedev groups N_0, M^+ and M^- fail to satisfy (*), and consequently the variety of ℓ-groups satisfying (*) does not contain any of the Medvedev covers. Thus, the variety of ℓ-groups satisfying (*) must contain a cover of $\mathcal{A}$ distinct from the Medvedev varieties, if this variety is in fact bigger than $\mathcal{A}$. To show this, suppose that G is a non-abelian o-group with the property that if $1 < b \ll a$, then $b \ll b^a$; such an order is called *special*. It is straightforward to show that an o-group with a special order satisfies (*). Thus, to find a cover of $\mathcal{A}$ contained in $\mathcal{R}$ distinct from the Medvedev varieties, it suffices to demonstrate the existence of an o-group with a special order (as observed by Feil [85]). This has been done by Bergman [84] who in fact proves that every free group can be given a special order; see the discussion on the variety *Spec* in the appendix. Exactly how many covers of $\mathcal{A}$ there are contained in $\mathcal{R}$ remains open. Note that recently Gurchenkov [84a] has shown that all proper varieties of ℓ-groups have covers in the lattice of varieties. In fact, except for $\mathcal{E}$ and $\mathcal{N}$, infinitely many; his proof makes use of the varieties $\mathcal{L}_n$. At present little is known about the structure of the covers of an arbitrary variety (other than $\mathcal{E}$, $\mathcal{A}$ and $\mathcal{N}$).

Section 7.3: *The cardinality of the lattice of ℓ-group varieties*

We now turn to the question of determining the cardinality of the lattice of ℓ-group varieties. Since each variety is determined by a set of equations and there are only countably many equations, the cardinality of the lattice of ℓ-group varieties can be no more than the cardinality of the continuum. We shall show that this maximum is attained. We will in fact show that the cardinality of both the height and breadth of the lattice is that of the continuum.

We shall first exhibit a tower of varieties whose cardinality is that of the continuum, following the construction of Feil [80]. Notice that we use additive notation.

Let w_1 and w_2 be two fixed ℓ-group words. For positive integers p and q, let $\mathcal{W}_{p/q}$ be the variety of representable ℓ-groups that satisfy $p|w_1| \geq q|w_2|$. (Hence $\mathcal{W}_{p/q}$ is the ℓ-group variety defined by $p|w_1| \geq q|w_2|$ and $2(x\wedge y) = 2x\wedge 2y$, by Theorem 4.1.1.) For representable ℓ-groups $nx = ny$ implies that $x = y$. Thus if $0 < \frac{p}{q} \leq \frac{m}{n}$, then $p|w_1| \geq q|w_2|$, which implies $mp|w_1| \geq mq|w_2| \geq np|w_2|$, which in turn implies that $m|w_1| \geq n|w_2|$. Thus $\mathcal{W}_{p/q} \subseteq \mathcal{W}_{m/n}$. It follows that if $\frac{m}{n} = \frac{p}{q}$ then $\mathcal{W}_{m/n} = \mathcal{W}_{p/q}$, and so these varieties are well-defined. Of course, for certain choices of w_1 and w_2 these inclusions may not be proper.

Now consider the words $w_1 = [x, |[x,y]|]$ and $w_2 = [x,y]$. For this choice of words and for any positive integers p and q with $0 < \frac{p}{q} \leq 1$, we define varieties $\mathcal{U}_{p/q}$ to be the varieties $\mathcal{W}_{p/q}$ as defined above. We shall now show that if $\frac{p}{q} < \frac{m}{n}$, then $\mathcal{U}_{p/q} \subset \mathcal{U}_{m/n}$.

For $t \in \mathbf{R}$, $0 < t \leq 1$, define G_t to be the group of ordered pairs (r,n), with $r \in \mathbf{R}$ and $n \in \mathbf{Z}$, with addition $(r,n)+(s,m) = (r+s(\frac{t}{t+1})^n, n+m)$. Then G_t is a splitting extension of $\mathbf{R}$ by $\mathbf{Z}$ with conjugation given by $-1_{\mathbf{Z}} + r + 1_{\mathbf{Z}} = \frac{r(t+1)}{t}$, where $1_{\mathbf{Z}}$ is the "one" of $\mathbf{Z}$. Each element of G_t can then be uniquely expressed as $r+n$ where $r \in \mathbf{R}$ and $n \in \mathbf{Z}$. We then order G_t lexicographically by calling an element positive if $n > 0$ or if $n = 0$ and $r > 0$, making G_t an o-group.

Lemma 7.3.1 For $0 < t \leq \frac{p}{q} \leq 1$, $G_t \in \mathcal{U}_{p/q}$.

Proof: Let $x, y \in G_t$ with $x \geq y \geq 0$. We shall show that x and y satisfy the defining inequality of $\mathcal{U}_{p/q}$, for $\frac{p}{q} \geq t$. We can write $x = r+n$, where without loss of generality $n > 0$ (else the result is trivial). Now $[x,y] \in \mathbf{R}$. Let $s = |[x,y]|$; we may as well assume that $s > 0$. Then

$$[x, |[x,y]|] = -n-r-s+r+n+s = -n-s+n+s = -s\left(\frac{t+1}{t}\right)^n + s = \left(1 - \left(\frac{1+t}{t}\right)^n\right)s.$$

Therefore

$$\big|[x, |[x,y]|]\big| = \left(\left(\frac{1+t}{t}\right)^n - 1\right)s \geq \left(\frac{1+t}{t} - 1\right)s = \frac{s}{t} = \frac{1}{t}|[x,y]|.$$

So if $\frac{p}{q} \geq t$ then $\frac{q}{p} \leq \frac{1}{t}$ and hence $\big|[x, |[x,y]|]\big| \geq \frac{q}{p}|[x,y]|$. It follows that $G_t \in \mathcal{U}_{p/q}$. //

Lemma 7.3.2 For $0 < \frac{p}{q} < t \leq 1$, $G_t \notin \mathcal{U}_{p/q}$.

Proof: Let $x = 1_{\mathbf{Z}}$ and $y = 1_{\mathbf{R}}$, the "ones" of $\mathbf{Z}$ and $\mathbf{R}$, respectively. Then $|[x,y]| = \frac{1}{t} \in \mathbf{R}$. So

$$[x, |[x,y]|] = -1_{\mathbf{Z}} - \frac{1}{t} + 1_{\mathbf{Z}} + \frac{1}{t} = -\left(\frac{1+t}{t}\right)\frac{1}{t} + \frac{1}{t} = \frac{-1}{t^2}.$$

So,

$$\big|[x, |[x,y]|]\big| = \frac{1}{t^2} < \frac{q}{pt} = \frac{q}{p}|[x,y]|.$$

It follows that $G_t \notin \mathcal{U}_{p/q}$. //

The previous two lemmas yield the following:

Theorem 7.3.3 $\{\mathcal{U}_{p/q} : 0 < \frac{p}{q} \leq 1\}$ is a tower of (representable) ℓ-group varieties and if $0 < \frac{m}{n} < \frac{p}{q}$, then $\mathcal{U}_{m/n} \subset \mathcal{U}_{p/q}$.

Now, for $r \in \mathbf{R}\backslash\mathbf{Q}$ and $0 < r < 1$, define $\mathcal{U}_r = \bigcap\{\mathcal{U}_{p/q} : \frac{p}{q} > r\}$. It follows that $\mathcal{U}_{p/q} \supset \mathcal{U}_r \supset \mathcal{U}_{m/n}$ for $\frac{p}{q} > r > \frac{m}{n}$. In fact, $G_r \in \mathcal{U}_r \backslash \mathcal{U}_{p/q}$, for $\frac{p}{q} < r$.

This now gives us a tower of varieties whose order type is that of the continuum. In fact, since the ℓ-groups $G_t \in \mathcal{A}^2 \cap \mathcal{R}$, there exists such a tower contained in $\mathcal{A}^2 \cap \mathcal{R}$; let $\bar{\mathcal{U}}_r = \mathcal{U}_r \cap \mathcal{A}^2 \cap \mathcal{R}$.

There is also a "sister" tower of varieties $\mathcal{U}^-_{p/q}$ defined by the equations

$$p\big|[x^{-1}, |[x,y]|]\big| \geq q|[x,y]|,$$

along with the corresponding $\mathcal{U}^-_r$ and $\bar{\mathcal{U}}^-_r$. (See Feil [80a] and Huss and Reilly [84].) Then $\mathcal{U}^-_r \subseteq \mathcal{U}^-_s$ if and only if $s \leq r$. The only cover of $\mathcal{A}$ contained in the intersection of the $\mathcal{U}^-_r$ is $\mathcal{M}^-$.

Huss [84a] has observed that the collection of varieties $\{\mathcal{U}_t \vee \mathcal{U}^-_t\}$ is pairwise incomparable, hence this is an antichain of varieties the size of the continuum.

We shall now describe Reilly's construction [81] of a continuum of ℓ-group varieties. For any group word u of the free group F on a countable set, let $l(u)$ be the law $z^+ \wedge u^{-1}z^-u = 1$. For any fully invariant subgroup U of F let $L(U)$ be the class of ℓ-groups satisfying the laws $l(u)$, for all $u \in U$. Let Θ be the usual map from the lattice of group varieties to the lattice of fully invariant subgroups of F (that is, $\mathcal{V}\Theta$ is the normal subgroup of F generated by the words defining the group variety $\mathcal{V}$). Reilly shows that the composition of Θ and L is a one-to-one complete lower semilattice homomorphism from the lattice of group varieties into the lattice of ℓ-group varieties. Furthermore, the ℓ-group varieties obtained under this map all contain $\mathcal{R}$. Adyan [70] has shown the existence of an antichain of group varieties of size of the continuum. Under Reilly's mapping those yield an antichain of ℓ-group varieties of size of the continuum.

From the two collections described above, we see that the breadth of the lattice of ℓ-group varieties has cardinality of the continuum.

Chapter 8: *Completions of Representable and Archimedean ℓ-groups*

Section 1: *Completions of representable ℓ-groups*

The idea of constructing a completion for a lattice-ordered group arises in two rather different contexts. On the one hand, it is natural to inquire as to the relationship between an ℓ-group and the ℓ-group inside of which it might be represented by one of the representation theorems discussed in Chapters 2 through 5. Can the target ℓ-groups for these representation theorems be described algebraically in terms of the ℓ-group to be represented? One might naturally expect such results to be phrased in terms of the larger ℓ-group being complete in some sense.

On the other hand, it is sometimes desirable to be able to embed a given ℓ-group into another with a given property, but in a minimal way. This process can give insight into how far the ℓ-group is from possessing the given property. Furthermore, it is often possible to prove theorems in the presence of a given property, and then prove the general result by restricting down from a completion.

We shall begin our discussion of completions in this section by considering the class of representable ℓ-groups (see Chapter 4); we use additive notation in this chapter. Every such ℓ-group can of course be embedded into a direct product of o-groups; a direct product of o-groups possesses most of the desirable completeness properties we shall discuss. However, the representation group is generally much bigger than the ℓ-group being represented, as we shall make clear in the discussion below. This makes impossible for most representable ℓ-groups the first aim of completion theory mentioned above.

We shall say that an ℓ-subgroup G of an ℓ-group H is *large* in H (alternatively, H is an *essential extension* of G) if every nonzero convex ℓ-subgroup of H has non-zero intersection with G. This is equivalent to asserting that for all $h \in H^+$, there exists $g \in G^+$ and positive integer n for which $g \leq nh$. This latter description of largeness suggests a slightly stronger relation between G and H : G is *order dense* in H if for all $h \in H^+$ there exists $g \in G$ with $0 < g \leq h$. We shall often merely say that G is *dense* in H; the reader should not infer any topological content in this terminology. Note that denseness does not mean that there exists an element of G between every comparable pair from H; see E17.

The origin of the term essential extension in module theory suggests immediately the possibility of maximal such extensions. However, if G is large (or dense) in H, and H is extended lexicographically by any totally ordered group whatsoever, G remains large (or dense) in the resulting ℓ-group. It is only by restricting consideration to archimedean ℓ-groups that the idea of maximal essential extension becomes fruitful, as we shall discover in the next section.

However, the idea of essential extension is the crucial one in making possible the construction of unique minimal ℓ-group extensions possessing certain desirable properties. We shall begin by showing that the Boolean algebra of polars remains the same in essential extensions.

We first introduce some notation in order to state this result more precisely. If G is an ℓ-subgroup of an ℓ-group H, $P \in \mathcal{P}(H)$ and $Q \in \mathcal{P}(G)$, and we denote the polar operations of G and H by $'$ and *, then we shall define $\Phi(P) = P \cap G$ and $\Psi(Q) = (Q')^*$.

Theorem 8.1.1 Let G be a large ℓ-subgroup of the ℓ-group H. Then the lattices $\mathcal{P}(G)$ and $\mathcal{P}(H)$ are isomorphic under the maps Φ and Ψ.

Proof: We first observe that if $Q \in \mathcal{P}(G)$, then $\Phi(\Psi(Q)) = (Q')^* \cap G = Q$, whenever G is an ℓ-subgroup of H.

For the opposite composition we shall prove that if $P \in \mathcal{P}(H)$, then $P \cap G = (P^* \cap G)'$. It is obvious that $P \cap G \subseteq (P^* \cap G)'$. If however, there exists $0 < g \in (P^* \cap G)' \backslash (P \cap G)$, then there exists $h \in P^*$ with $g \wedge h > 0$. Because G is large in H, there exists $0 < k \in G \cap H(g \wedge h)$. But $h \in P^*$ and so k is too, which means that $k \wedge g = 0$, which is a contradiction. Thus

$$\Psi(\Phi(P)) = \Psi(P \cap G) = \Psi((P^* \cap G)') = (P^* \cap G)^* \supseteq P.$$

If $0 < b \in (P^* \cap G)^* \backslash P$, then $b \wedge a > 0$ for some $a \in P^*$, and hence there exists $c \in G^+ \cap H(b \wedge a)$. But then $c \in P^*$ and so $b \wedge c = 0$, a contradiction. Thus $\Psi(\Phi(P)) = P$.//

A useful property possessed by essential extensions is that they preserve all (infinite) meets and joins of the large ℓ-subgroup. An easy proof exists for this if the ℓ-groups are abelian (see Bernau [65b]); the following proof is Bleier and Conrad's [75]:

Theorem 8.1.2 Suppose G is a large ℓ-subgroup of the ℓ-group H and A is a subset of G which has least upper bound g in G. Then g is the least upper bound of A in H as well. The dual statement holds for meets.

Proof: Suppose to the contrary that there exists $h \in H$ with $g > h \geq a$, for all $a \in A$. Because G is large in H, there exists a positive integer n and $x \in G$ such that $n(g - h) \geq x > 0$. Then,

$$0 \geq x + n(h - g) \geq x + \sum_{i=1}^{n}(a_i - g),$$

for any collection $\{a_i\}$ of n elements from A. But then,

$$-(x + \sum_{i=1}^{n-1}(a_i - g)) + g \geq a_n,$$

for all $a_n \in A$. Since the elements to the left of the inequality all belong to G, we then have that

$$-(x + \sum_{i=1}^{n-1}(a_i - g)) + g \geq g, \text{ or } 0 \geq x + \sum_{i=1}^{n-1}(a_i - g).$$

We can then repeat this argument $n - 1$ times, deleting one term of the sum at a time, until we arrive at $0 \geq x$, which is a contradiction. //

We shall now make some definitions, preparatory to discussing the most important of the completions of a representable ℓ-group, the orthocompletion. An ℓ-group is *strongly projectable* if it has as many cardinal summands as possible; namely, all polars are summands (see the discussion following Theorem 1.2.5). As mentioned at the end of chapter 5, an ℓ-group is *laterally complete* if each pairwise disjoint set of elements has a least upper bound. An ℓ-group with both of these properties is said to be *orthocomplete*. Notice that a strongly projectable ℓ-group (and hence an orthocomplete one) is necessarily representable, since each polar, as a summand, is normal (Theorem 4.1.1).

An *orthocompletion* H of an ℓ-group G is a minimal essential extension of G which is orthocomplete. (It would not alter what follows were we to require that G be dense in H instead.) Also, since ℓ-subgroups of representable ℓ-groups are representable (since the class of representable ℓ-groups is a variety), it only makes sense to speak of orthocompletions of representable ℓ-groups. Ball [82] has generalized these notions to the non-representable case. We shall prove the existence and uniqueness of the orthocompletion, for any representable ℓ-group.

Such questions were first considered for vector lattices by Nakano [50] and Amemiya [53]; Bernau [66b] gave the original construction for arbitrary representable ℓ-groups (which he called 'orthocommutative'). His construction involves a laborious identification of the positive cone of the orthocompletion in terms of equivalence classes of subsets of the ℓ-group; Conrad ([69],[73]) showed that this construction is essentially a direct limit of cardinal products of quotients by polars. We shall here present Bleier's construction [76] of the orthocompletion. Much different than Conrad's, it has the advantage of relying on the Lorenzen representation and is consequently more intuitively satisfying.

In order to prove the existence of an orthocompletion for a representable ℓ-group G, we shall assume a Lorenzen representation

$$G \longrightarrow \prod\{T_\lambda : \lambda \in \Lambda\},$$

together with the isomorphic Boolean algebras $\mathcal{P}(G)$ and

$$\mathcal{P}(\Lambda) = \{spt(P) : P \in \mathcal{P}(G)\},$$

as discussed in Section 4.1.

We need a few definitions from the theory of Boolean algebras. A *partition of an element* is a maximal disjoint collection of elements below it; by a *partition* we shall understand a partition of the maximum element. A partition A *refines* B if every element of A is below some element of B. It is evident that any two partitions admit a common refinement, consisting of all non-zero intersections of elements from A with elements of B.

For $a \in \Pi T_\lambda$, suppose that there exists a partition $\{P_\alpha : \alpha \in A\}$ in $\mathcal{P}(\Lambda)$, and a set of elements $\{g_\alpha : \alpha \in A\}$ of G, so that $a(\lambda) = g_\alpha(\lambda)$, for all $\lambda \in P_\alpha$. We shall say that $\{g_\alpha, P_\alpha\}$ *underlies* a. We can express this intuitively by saying that a can be piecewise approximated by elements of G. Now let L be the set of all elements of ΠT_λ for which such a partition and such a subset of G exists.

By taking common refinements where necessary, it is easily proved that L is an ℓ-subgroup of ΠT_λ. This will provide us with an orthocompletion of G once we stop distinguishing among elements of L which disagree on topologically insignificant sets. More precisely, for $f, h \in L$, we define $f \sim h$ to mean that there exist a partition $\{P_\alpha\}$ and a subset $\{g_\alpha\}$ of G so that $\{g_\alpha, P_\alpha\}$ underlies both f and h. It is easily verified that this is an ℓ-group congruence on L; the homomorphic image we shall call $O(G)$. Note that as an ℓ-homomorphic image of an ℓ-subgroup of ΠT_λ, it is evident that $O(G)$ is at least representable. We shall prove that $O(G)$ is an orthocompletion of G.

Given equivalence classes $[f], [g] \in O(G)$, it is straightforward to verify that $[f] < [g]$ means just that f is "essentially" below g: that is, if $\{f_\alpha, F_\alpha\}$ underlies f, then there exists $\{g_\beta, G_\beta\}$ underlying g, so that $f_\alpha(\lambda) \leq g_\beta(\lambda)$, for all $\lambda \in F_\alpha \cap G_\beta$.

This observation makes it easy to show that $O(G)$ contains a copy of G, under the map $g \mapsto [g]$. For if g is a nonzero element of G, but $g \sim 0$, then there exists a collection $\{g_\alpha, P_\alpha\}$ underlying both g and 0 . Since $\{P_\alpha\}$ is a partition, $g'' \cap P_\alpha$ is not zero, for some α. This means that there exists $\lambda \in spt(g'') \cap spt(P_\alpha)$. But then $g(\lambda) = g_\alpha(\lambda) = 0(\lambda) = 0$, which is a contradiction.

Furthermore, this copy of G is dense in H. For if $[h]$ is a positive element of $O(G)$, we may assume that underlying it is $\{h_\alpha, H_\alpha\}$, with $h_\alpha \geq 0$, for all α. If $[h]$ is not zero, then there exists α for which $h''_\alpha \cap H_\alpha \neq \{0\}$. Consequently, if we choose $g \in h''_\alpha \cap H_\alpha$, then $[g \wedge h_\alpha]$ is the required nonzero element of G which lies below $[h]$. We shall henceforth identify G with its image $[G]$ under this map.

Since G is large in $O(G)$, we know from Theorem 8.1.1 that their Boolean algebras $\mathcal{P}(G)$ and $\mathcal{P}(O(G))$ of polars are isomorphic; also, $\mathcal{P}(G)$ is isomorphic to $\mathcal{P}(\Lambda)$. Putting these isomorphisms together means that $Q \in \mathcal{P}(O(G))$ corresponds to $spt(Q \cap G)$. Now suppose that $\{h_\alpha, H_\alpha\}$ underlies h in L. Define $k(\lambda) = h_\alpha(\lambda)$ if $\lambda \in H_\alpha \cap spt(Q \cap G)$ and zero otherwise. Since $\{H_\alpha\}$ is a partition, this makes k a well-defined element of L. Clearly, $[k]$ is an element of Q. We can then similarly define an element $[g]$ of Q'; we then have $[h] = [k + g] = [k] + [g]$. Thus, $O(G)$ is strongly projectable.

To complete the proof that $O(G)$ is orthocomplete, we need to show that each pairwise disjoint set in $O(G)$ has a least upper bound. Toward that end, suppose that $\{[h_\beta]\}$ is a pairwise disjoint set of elements in $O(G)$. We may then suppose that the supports of the elements h_β of L are pairwise disjoint, and that each h_β is underlain by $\{h_{\beta\alpha}, P_{\beta\alpha}\}$ together with 0 paired with Q_β, where Q_β is the element of $\mathcal{P}(\Lambda)$ corresponding to the polar h^*_β of $O(G)$. Because the P's are all pairwise disjoint, we may define $h \in \Pi T_\Lambda$ by setting $h(\lambda) = h_{\beta\alpha}(\lambda)$ if $\lambda \in P_{\beta\alpha}$, and 0 otherwise. Furthermore, $\{h_{\beta\alpha}, P_{\beta\alpha} : \text{all } \alpha \text{ and } \beta\}$, together with the element $Q \in \mathcal{P}(\Lambda)$ corresponding to $\{h_\beta\}^{**}$ paired with 0 , underlies h, and so $h \in L$. It is obvious that $h_\beta \leq h$, for all β, and so $[h_\beta] \leq [h]$. Now, if $[h_\beta] \leq [f]$, for all β. we can choose a refinement $\{f_\gamma, R_\gamma\}$ of the partition $\{P_{\beta\alpha}\} \cup \{\{Q\}\}$ underlying f, so that on each R_γ, $h_\beta \leq f_\gamma$ (for the unique h_β which is nonzero on R_γ). This makes it clear that $[h] \leq [f]$.

We have now shown that G can be ℓ-embedded as a large ℓ-subgroup of an orthocomplete ℓ-group $O(G)$. To show that $O(G)$ is an orthocompletion of G, it remains to prove that if H is an orthocomplete ℓ-subgroup of $O(G)$ which contains G, then $H = O(G)$. But if $h \in O(G)^+$, we may choose $\{h_\beta, P_\beta\}$ which underlies h, and so h is the least upper bound of $\{h_\beta[P_\beta]\}$, where $h_\beta[P_\beta]$ is the projection in H of h_β on the polar corresponding to P_β. Since H is orthocomplete, this means that h is an element of H.

It remains to prove that any other orthocompletion of G is ℓ-isomorphic to $O(G)$. This is accomplished by showing that if H is any other orthocomplete essential extension of G, then $O(G)$ can be ℓ-embedded into H by a map which leaves G fixed. Given a positive element $[h] \in O(G)$, it is underlain by $\{h_\beta, P_\beta\}$, and so as above $[h] = \bigvee\{h_\beta[P_\beta]\}$. Each set P_β corresponds to a polar Q_β of H, since G is large in H. We can then define a function ϕ from $O(G)^+$ to H^+ by setting $\phi([h]) = \bigvee\{h_\beta[Q_\beta]\}$. By taking suitable refinements one can show that this is well-defined. This function clearly leaves elements of G fixed; one can check that it is an ℓ-monomorphism.

We have thus proved

Theorem 8.1.3 Every representable ℓ-group admits a unique orthocompletion.

We observe here for future reference that it follows immediately from the construction of the orthocompletion of ℓ-group G that each positive element of it admits a representation of the form $\bigvee\{g_\alpha[P_\alpha]\}$, where each g_α is a positive element of G, and $g_\alpha[P_\alpha]$ is the projection of g_α onto polar P_α, where $\{P_\alpha\}$ is a partition in $\mathcal{P}(O(G))$.

Obviously, an ℓ-group coincides with its orthocompletion exactly if it is already orthocomplete. At the other extreme, it coincides with the entire group $\prod T_\lambda$, for some Lorenzen representation, exactly if it contains the small sum $\sum T_\lambda$ of the "stalks" T_λ and so has a basis (defined in section 4.1).

Once the orthocompletion has been constructed, it is relatively easy to demonstrate the existence and uniqueness of various other smaller completions for representable ℓ-groups. For example, let $X = SP$, P or L; by an SP-group we mean an ℓ-group which is strongly projectable (see class $\mathcal{SP}$ in the appendix), by a P-group one for which all polars of the form g'' are summands (projectable, see class $\mathcal{P}$ in the appendix), and by an L-group, one which is laterally complete. Since $O(G)$ is both laterally complete and strongly projectable, the intersection of all X-subgroups of $O(G)$ containing G is clearly a minimal essential extension of G which is an X-group. It is fairly easy to show that such hulls are unique (see Conrad [73]).

The crucial ingredient for the proofs of this chapter is a representation as a subdirect product where disjointness corresponds to disjointness of support; indeed, as demonstrated in Conrad [75b] and [78], the same proofs apply to completions of semiprime rings, considered as subdirect products of prime rings. An interesting case where both of these structures obtain, namely where the ℓ-group (or ring) is the set of continuous real-valued functions on a topological space, has been considered in some detail in Anderson and Conrad [82].

In the last mentioned situation, the ℓ-groups considered are archimedean. Application of the orthocompletion to this case will occupy us in the next section.

Section 8.2: *Completions of archimedean ℓ-groups*

In this section we shall apply the results obtained in the previous section regarding completions of representable ℓ-groups to the archimedean case. In the process we shall show that the target ℓ-group $D(X)$ for Bernau's representation theorem is uniquely determined algebraically by the ℓ-group so represented; we shall also consider the ℓ-group analogue of the Dedekind construction of the reals from the rationals.

We shall begin by returning to the notion of largeness, in the context of archimedean ℓ-groups. An archimedean ℓ-group is *essentially closed* if it admits no proper archimedean essential extensions. An *essential closure* of an archimedean ℓ-group is an essential extension of it which is essentially closed. As we remarked at the beginning of the last section, these definitions are vacuous when considered in the context of all ℓ-groups. In light of Bernau's representation theorem for archimedean ℓ-groups, and the fact that a complete Boolean algebra uniquely determines its Stone space, the following theorem, first proved by Conrad [71a], seems immediately plausible:

Theorem 8.2.1 Each archimedean ℓ-group G admits a unique essential closure G^e, which is of the form $D(X)$, where X is the Stone space corresponding to $\mathcal{P}(G)$.

Proof: We first show that ℓ-groups of the form $D(X)$ are essentially closed. Suppose that H is an archimedean essential extension of $D(X)$. Since $\mathcal{P}(H)$ is isomorphic to $\mathcal{P}(D(X))$, there exists an ℓ-embedding of H into $D(X)$. We may choose this ℓ-embedding by letting the single constant function $\bar{1}$ in $D(X) \subseteq H$ serve as the maximal pairwise disjoint set in the proof of Bernau's representation theorem (Theorem 2.3). But then the composition of these ℓ-embeddings is a map from $D(X)$ into $D(X)$ which takes $\bar{1}$ to $\bar{1}$; by examining the proof of Bernau's representation theorem we see that this map must be the identity, and so $H = D(X)$.

It is now obvious that each archimedean ℓ-group admits an essential closure of the form $D(X)$. The uniqueness follows because any other essential extension could in turn be ℓ-embedded into $D(X)$ for the same X, since the Boolean algebra of polars is the same. //

This theorem accomplishes for Bernau's representation theorem the first of the aims for completion theorems discussed at the beginning of this chapter. But to better understand how far from an archimedean ℓ-group its essential closure might be, we must examine two particular completions of archimedean ℓ-groups, which are of independent interest.

It is natural to ask whether it is possible to complete an arbitrary ℓ-group in a way directly analogous to Dedekind's construction of the real numbers from the rationals. Toward that end, we make the obvious definition of a *complete* ℓ-group: namely, by specifying that each set of elements which is bounded above has a least upper bound. Note that this implies dually that each set bounded below has a greatest lower bound. Furthermore, a complete ℓ-group is archimedean, for if $0 \le ng < h$, for all positive n, then $k = \vee ng$ exists. But then

$$g + k = g + \vee ng = \vee ng = k,$$

which implies that $g = 0$. Consequently it only makes sense to talk about a Dedekind completion of an archimedean ℓ-group(using the topological methods we discuss in Chapter 9 Ball [80] generalizes this to nonarchimedean ℓ-groups).

Following again the example of the construction of **R** from **Q**, we shall call $G^\wedge$ a *Dedekind completion* of an ℓ-group G if G is an ℓ-subgroup of $G^\wedge$, $G^\wedge$ is complete, and each element of $G^\wedge$ is the least upper bound of a subset of G (again, the dual statement follows from this).

It is possible to obtain $G^\wedge$ from G by constructing the cut completion of G as a partially ordered set, following MacNeille's generalization [37] of Dedekind's original method, and then defining an addition on an appropriate subset to create $G^\wedge$, as done by Everett and Ulam [45]. We shall not follow this procedure, but shall instead identify $G^\wedge$ as an ℓ-subgroup of G^e, following the lead of Conrad and McAlister [69].

To do this we need two facts: first, that all meets and joins in a dense ℓ-subgroup are preserved in the larger ℓ-group (Theorem 8.1.2), and second, that any essentially closed archimedean ℓ-group is complete. For the second statement, consider that Nakano [41e] and Stone[40] have shown that if X is any extremally disconnected space, then $C(X)$, the ℓ-group of real-valued continuous functions on X, is complete. From this, it follows easily that ℓ-groups of the form $D(X)$ are complete, because $C(X)$ is dense in $D(X)$, and for any $g \in D(X)^+$ with $g(x) > 0$ for all x, the convex ℓ-subgroup of $D(X)$ generated by g is evidently ℓ-isomorphic to $C(X)$ (via multiplication by $\frac{1}{g}$).

We now claim that for any archimedean ℓ-group G,

$$G^\wedge = \{h \in G^e : h = \vee\{g \in G : g \leq h\}\}$$

is a completion of G in the sense of the definition above. We first must verify that $G^\wedge$ is in fact a group. For that purpose, suppose that $a, b \in G^\wedge$, and so $a = \bigvee A$ and $b = \bigvee B$, where $A = \{g \in G : g \leq a\}$ and $B = \{g \in G : g \leq b\}$. Now let $C = \{g \in G : g \leq a - b\}$. Since G^e is complete, there certainly exists $c \in C$ with $c = \bigvee C$. We claim that $c = a - b$. Clearly $c \leq a - b$ and so $c + b \leq a$. This means that $g + b \leq a$ for all $g \in C$. But then the coset $C + b \subseteq A$. On the other hand, if $h \in A$, then $h - b \leq a - b$ and so $h \in C + b$. Thus $C + b = A$ and so

$$c + b = \bigvee(C + b) = \bigvee A = a;$$

that is, $c = a - b$, as required.

The reader may verify that this set in fact forms an ℓ-subgroup of G^e; it is then evident that it is a completion. We shall then have completed the proof of the following theorem when we check for uniqueness:

Theorem 8.2.2 Each archimedean ℓ-group admits a unique Dedekind completion.

Proof: Let G be an archimedean ℓ-group. To complete the proof one need merely verify that if ι is an ℓ-embedding of G into H making H another completion of G, we may define a map ϕ from $G\hat{}$ into H given by

$$\phi(k) = \bigvee_H \{\iota(g) : g \leq k\}.$$

It is easily checked that this is an ℓ-isomorphism. //

Note that of course G is dense in $G\hat{}$ and that if we wish we may assume that $G\hat{}$ is an ℓ-subgroup of G^e.

Before leaving the topic of Dedekind completeness we must prove the following important property of complete ℓ-groups, first proved by F. Reisz [40]:

Theorem 8.2.3 A complete ℓ-group is strongly projectable.

Proof: Suppose that G is a complete ℓ-group and P is a polar of G; we must show that P is a cardinal summand of G. Let $g \in G^+$. Since G is complete we can define $g[P] = \vee\{k \in P : k \leq g\}$. Since it is clear that $g[P] \in P$ it remains only to show that $g - g[P] \in P'$. But if $h \in P^+$, then

$$h \wedge (g - g[P]) = ((g[P] + h) \wedge g) - g[P].$$

Now, $(g[P] + h) \wedge g \in P$, and since it is less than or equal to g, it is less than or equal to $g[P]$. This means that $h \wedge (g - g[P]) = 0$; thus $g - g[P] \in P'$. //

We now shall consider the orthocompletion of an archimedean ℓ-group. It is surprisingly the case that this is just the lateral completion, because lateral completeness also implies strong projectability, in the presence of the archimedean hypothesis. (See E35 for a non-archimedean laterally complete ℓ-group which is not strongly projectable.) The proof of this implication is much more difficult than that of Theorem 8.2.3. Veksler and Geiler [72] first obtained this result, in the context of vector lattices; their proof makes use of the representation in $D(X)$. The following elementary proof is due to Bernau [75b].

Theorem 8.2.4 An archimedean laterally complete ℓ-group is strongly projectable.

Proof: Let G be an archimedean laterally complete ℓ-group. We first observe that it is sufficient to prove that G is projectable. For if it is, let P be any polar of G. Then $P = \sqcup g_\alpha''$ for some maximal pairwise disjoint subset $\{g_\alpha\}$ of P^+. Then for any $h \in G^+$, the projections $h[g_\alpha'']$ all exist, and so does $\vee h[g_\alpha'']$, since these elements are pairwise disjoint. It is easily seen that this is just $h[P]$.

Suppose now that $g, h \in G^+$. We shall show that h is an element of $g'' \boxplus g'$. For each nonnegative integer n , let

$$w_n = ((n+2)g - h)^+ \wedge (h - ng)^+.$$

It is obvious that each $w_n \in g''$, and that for $k \geq 2$, $w_n \wedge w_{n+k} = 0$. Thus since G is laterally complete, we have the existence of

$$u = \bigvee\{(2n+1)w_{2n} : n = 0, 1, 2, \ldots\} \text{ and}$$

$$v = \bigvee\{(2n+2)w_{2n+1} : n = 0,1,2,\ldots\}.$$

We shall show that $h[g''] = h \wedge (u+v)$. Since it is evident that $h \wedge (u+v) \in g''$, it remains to prove that $h - h \wedge (u+v)$ is an element of g'.

Now, the inequalities

$$\begin{aligned}
&\left[(n+1)((n+2)g-h)^+\right] \wedge h + (n+2)w_{n+1} \\
&\geq \left[(n+1)((n+2)g-h)^+ + (n+2)w_{n+1}\right] \wedge h \\
&= \left[(n+1)((n+2)g-h)^+ + (n+2)\{((n+3)g-h)^+ \wedge (h-(n+1)g)^+\}\right] \wedge h \\
&\geq \left[(n+2)((n+3)g-h)^+\right] \wedge \left[(n+1)((n+2)g-h) + (n+2)(h-(n+1)g)\right] \wedge h \\
&= \left[(n+2)((n+3)g-h)^+\right] \wedge h,
\end{aligned}$$

give an inductive proof that

$$\sum_{r=0}^{n}(r+1)w_r \geq \left[(n+1)((n+2)g-h)^+\right] \wedge h \geq (ng-h)^+ \wedge h.$$

Thus for each nonnegative integer k,

$$\begin{aligned}
g \wedge (h - h \wedge (u+v)) &\leq g \wedge (h - h \wedge \sum_{r=0}^{n}(r+1)w_r) \\
\leq g \wedge (h - h \wedge (kg-h)^+) &\leq g \wedge (h - h \wedge (kg-h)) \\
= g \wedge (2h - 2h \wedge kg) &= ((g+2h) \wedge (g+kg) \wedge 2h) - (2h \wedge kg) \\
&= 2h \wedge (k+1)g - 2h \wedge kg.
\end{aligned}$$

Therefore for any positive integer n

$$n\left[g \wedge (h - h \wedge (u+v))\right] \leq \sum_{k=0}^{n-1}[2h \wedge (k+1)g - 2h \wedge kg] \leq 2h.$$

Since G is archimedean, we have $h - h \wedge (u+v) \in g'$, as required. //

We can now use this theorem to explicitly determine the orthocompletion of an archimedean ℓ-group. We first observe that if X is a Stone space, then $D(X)$ is laterally complete. For, if $\{g_\alpha\}$ is a pairwise disjoint subset of $D(X)^+$, then as functions on X their supports are pairwise disjoint open subsets of X. We can now define a function g by letting $g(x) = g_\alpha(x)$, if $x \in spt(g_\alpha)$, and by letting it be zero on the support of $\{g_\alpha : \text{all } \alpha\}'$. We now have defined g continuously on an open dense subset of X, and consequently it admits a unique extension to all of X which is evidently the join of the set $\{g_\alpha\}$ in $D(X)$. Note that a slight modification of this argument proves that each element of $D(X)^+$ is in fact the join of a pairwise disjoint subset of $C(X)^+$.

Because each $D(X)$ is laterally complete, any archimedean ℓ-group G can be ℓ-embedded as a large subgroup of an archimedean laterally complete (and hence orthocomplete) ℓ-group, namely some $D(X)$. But then the intersection of all orthocomplete

ℓ-subgroups of $D(X)$ containing G is clearly an orthocompletion for G. Since orthocompletions are unique it follows that the orthocompletion (or equivalently, lateral completion) of an archimedean ℓ-group is archimedean, and can be considered an ℓ-subgroup of G^e. We summarize this, and prove a bit more, in the following theorem:

Theorem 8.2.5 Each archimedean ℓ-group G admits a unique lateral completion G^L, which is archimedean and equal to the orthocompletion of G. Furthermore G^L consists of the subgroup of G^e generated by joins of disjoint subsets of G.

Proof: It remains only to prove the last statement of the theorem. For that purpose, suppose that G is a large ℓ-subgroup of its lateral completion G^L.

We first strengthen a part of the argument in the proof of the last theorem: Suppose that $h, g \in G^{L+}$ and $h \in g''$; we claim that $h = u + v$, where u and v are the elements defined in the previous proof. Clearly, we need only check that $h \geq u + v$. Now

$$\begin{aligned}(n+1)w_n &= [h + (n+2)((n+1)g - h)]^+ \wedge [h + n(h - (n+1)g)]^+ \\ &= [h + [(n+2)((n+1)g - h) \wedge n(h - (n+1)g)]]^+ \leq h.\end{aligned}$$

Since

$$u + v = \bigvee\{(2n+2)w_{2n+1} + (2m+1)w2_m) : \text{all } m, n\},$$

and $w_n \wedge w_{n+p} = 0$ for all $p \geq 2$, it is enough to show that $(n+1)w_n + (n+2)w_{n+1} \leq h$, for nonnegative integers n. Since $a^+ + b^+ = (a+b)^+ \vee a^+ \vee b^+$, we only have to show that

$$(n+1)[((n+2)g - h) \wedge (h - ng)] + (n+2)[((n+3)g - h) \wedge (h - (n+1)g)]$$

is less than or equal to h. However, this last quantity is less than or equal to

$$(n+1)((n+2)g - h) + (n+2)(h - (n+1)g) = h.$$

Now suppose that $w \in G^{L+}$. Since G^L is the orthocompletion as well, there exists a partition of polars $\{P_\alpha\}$ and a collection of positive elements $\{g_\alpha\}$ of G so that $w = \vee\{g_\alpha[P_\alpha]\}$; we may as well assume that each P_α is of the form x''_α, where $x_\alpha \in G^+$. But by applying the argument of the previous paragraph to $g = x_\alpha$ and $h = g_\alpha[P_\alpha]$, we can obtain that $g_\alpha[P_\alpha] = u_\alpha + v_\alpha$, where u_α and v_α are joins of multiples of elements of the form

$$\begin{aligned}((n+2)x_\alpha - g_\alpha[P_\alpha])^+ \wedge (g_\alpha[P_\alpha] - nx_\alpha)^+ &= (((n+2)x_\alpha - g_\alpha)^+ \wedge (g_\alpha - nx_\alpha)^+)[P_\alpha] \\ &= ((n+2)x_\alpha - g_\alpha)^+ \wedge (g_\alpha - nx_\alpha)^+.\end{aligned}$$

But since elements of the latter form are pairwise disjoint for distinct α we can define elements u and v, by taking the appropriate joins for all α; consequently $w = u + v$, which is what we wanted to show. //

The proof above, which is due to Bernau, appeared in Anderson, Conrad and Kenny [77].

For non-archimedean ℓ-groups, it is far from the case that G^L is the group generated by joins of pairwise disjoint sets, as Bixler [85] (see E40) has demonstrated.

A corollary of the fact that laterally complete archimedean ℓ-groups are projectable is the following:

Theorem 8.2.6 For an archimedean ℓ-group G, $G^{\wedge L} = G^{L\wedge}$.

Proof: Since $G^\wedge$ is archimedean so is $G^{\wedge L}$, and thus is orthocomplete by Theorem 8.2.4. Since G is dense in all of the groups considered here, this means that $G^L \subseteq G^{\wedge L}$. To show that $G^{L\wedge} \subseteq G^{\wedge L}$ we must prove that $G^{\wedge L}$ is complete.

We first check that $G^\wedge$ is a convex ℓ-subgroup of $G^{\wedge L}$. Suppose that $0 < g < h$, with $g \in G^{\wedge L}$ and $h \in G^\wedge$. Since $G^\wedge$ is projectable, $g = \vee\{x_\alpha\}$, where $\{x_\alpha\}$ is a pairwise disjoint subset of $G^\wedge$. But then $\{x_\alpha\}$ is a bounded subset (by h) of $G^\wedge$ and so its join belongs to $G^\wedge$.

To show that $G^{\wedge L}$ is complete, we now suppose that $\{g_\alpha\}$ is a subset of $G^{\wedge L}$ which is bounded by h, which we may suppose is the join of the pairwise disjoint subset $\{y_\beta\}$ of $G^\wedge$. Then $\{g_\alpha \wedge y_\beta : \alpha\}$ is a bounded subset of $G^\wedge$ (since $G^\wedge$ is convex in $G^{\wedge L}$), and so has a least upper bound z_β. Clearly $\{z_\beta : \beta\}$ is a pairwise disjoint subset of $G^\wedge$, and so has a join $z \in G^{\wedge L}$. It is straightforward to verify that z is also the join of $\{g_\alpha\}$.

We now must show the opposite inclusion, namely that $G^{\wedge L} \subseteq G^{L\wedge}$. Since G is dense in $G^{L\wedge}$ and $G^{L\wedge}$ is complete, it is evident that $G^\wedge \subseteq G^{L\wedge}$. To complete the proof, we evidently need only verify that $G^{L\wedge}$ is laterally complete. For that purpose, suppose that $\{x_\lambda\}$ is a pairwise disjoint subset of $G^{L\wedge}$. For each λ choose $y_\lambda \in G^L$ with $y_\lambda \geq x_\lambda$. Since G^L is strongly projectable, we may as well assume that y_λ is an element of the polar $x_\lambda'' \cap G^L$. But then $\{y_\lambda\}$ is a pairwise disjoint subset of G^L, and as such has a least upper bound $y \in G^L$. Now, $\{x_\lambda\}$ is a bounded subset of $G^{L\wedge}$ and as such has a least upper bound. //

Conrad [73] proved that $G^{\wedge O} = G^{O\wedge}$; Bernau [75b] then obtained the theorem above by proving Theorem 8.2.4.

We are now ready to obtain the essential closure of an archimedean ℓ-group by application of the completions discussed above:

Theorem 8.2.7 Let G be an archimedean ℓ-group. Then $G^e = ((G^d)^\wedge)^L = ((G^d)^L)^\wedge$.

Proof: By the previous theorem we evidently need only prove the first of these equalities. Furthermore since G^e is of the form $D(X)$ where X is a Stone space, and such groups are divisible, complete, and laterally complete, we evidently have that $G^{d\wedge L} \subseteq G^e$.

To prove that this inclusion is in fact equality, we shall first show that $G^{d\wedge}$ is convex in G^e. For suppose that $0 < g < h$, with $g \in G^e$ and $h \in G^{d\wedge}$. Since G^d is dense in G^e, this means that $g = \vee\{k \in G^d : k \leq g\}$. But all such k are bounded by h, an element of $G^{d\wedge}$, and so this join exists in $G^{d\wedge}$; that is $g \in G^{d\wedge}$.

We shall have completed the proof if we show that the lateral completion of a dense convex ℓ-subgroup H of $D(X)$ is necessarily $D(X)$; we shall in fact show that each positive element of $D(X)$ is the join of a pairwise disjoint subset of H. We have already observed that this is case for the dense convex ℓ-subgroup $C(X)$. So we need only verify that if $f \in C(X)^+ \backslash H$, then f is the join of a pairwise disjoint subset of H. Multiplication by an appropriate element of $D(X)^+$ allows us to assume that f is the characteristic function on some clopen subset A of X. Let S be a maximal pairwise disjoint subset of H whose supremum is a characteristic function on some subset B of A. We would like to prove that $B = A$. If not, because H is dense in $D(X)$ there exists $g \in H^+$ which is less than the characteristic function on $A \backslash B$. There exists positive integer n for which open set $X_n = \{x \in X : g(x) > \frac{1}{n}\}$ is nonempty; we can find $k \in C(X)$ which agrees with g on X_n and is zero on the interior of $X \backslash X_n$. Since H is convex, $k \in H$, and consequently so is the meet of nk with the characteristic function on $A \backslash B$. But this latter function is a characteristic function, which contradicts the maximality of S. //

This theorem was proved by Conrad [71a]. We note here that it is essential to first "divisibilize" G in order that it be dense, rather than merely large, in its essential closure, as E22 reveals.

Chapter 9: *The Lateral Completion*

The question of the existence of a lateral completion for an arbitrary ℓ-group has been an important and fruitful one in the theory of ℓ-groups. Conrad's important paper [69] laid the groundwork for the problem, and proved existence and uniqueness for representable ℓ-groups and ℓ-groups with zero radical. Byrd and Lloyd [69] applied Conrad's methods to the completely distributive case (for a definition and results concerning complete distributivity, the reader should consult Chapter 10). It was Bernau [74b],[75a] who provided the first proof for arbitrary ℓ-groups. His construction, like that he made for the orthocompletion [66b], involves a very technical brute force construction of the elements of the positive cone of the lateral completion in terms of the original ℓ-group. This construction has since been simplified by McCleary [81], but it remains exceedingly complicated.

It might be expected that the lateral completion of an arbitrary ℓ-group could be found inside a Holland representation. The difficulty, as easy examples reveal (see E45), is that a given Holland representation for an ℓ-group, although laterally complete, need not contain a lateral completion of the represented ℓ-group. The permutation approach has in fact been successful, but only in the completely distributive case, where a Holland representation is available where all infinite suprema and infima are preserved; this work is due to Davis and McCleary [81a].

It was Rick Ball [80a] who first showed the existence and uniqueness of the lateral completion of an arbitrary ℓ-group by topological completion methods. Furthermore, his approach is of more general interest in the theory of ℓ-group completions, and consequently we will present here his construction of the lateral completion as an example of his methods.

The obvious approach to putting a topology on an ℓ-group in such a way as to make both the group and lattice operations continuous, is to let the set of open intervals serve as a base for the topology, as is done in the totally ordered case [KK]. Unfortunately, if the ℓ-group in question is not totally ordered, this merely leads to the discrete topology. Topologies can be put on ℓ-groups, as work by Redfield [74] , Conrad [74d], Madell [69], Smarda [67] and Kenny [75] demonstrates; the most important and comprehensive paper in this direction was written by Ball [79].

However, it was Ball [80a] who proved that more general convergence structures than those provided by topologies on the ℓ-group are necessary in order to obtain powerful completion results. In this chapter we describe these more general structures, in sufficient generality to achieve our goal of constructing the lateral completion.

We shall use multiplicative notation for ℓ-groups in this chapter.

Given subsets X and Y of an ℓ-group G, let

$$XY = \{xy : x \in X, y \in Y\}, \text{ and } X^{-1} = \{x^{-1} : x \in X\}.$$

The sets $X \wedge Y$ and $X \vee Y$ can be defined similarly.

The *convexification* $con(X)$ of X is defined as

$$con(X) = \{y : x \leq y \leq z, \text{ for } x, z \in X\}.$$

The *order closure*, $ocl(X)$, of X is defined as the smallest subset of G containing X which is closed under all (perhaps infinite) suprema and infima that exist in G.

A collection $\mathcal{F}$ of subsets of an ℓ-group G is a *filter base* if it is closed under finite intersections; it is a *filter* if it is closed also under taking supersets. Each filter base $\mathcal{F}$ uniquely determines a filter $\{X : X \supseteq Y, \text{some } Y \in \mathcal{F}\}$. Given filters $\mathcal{F}$, $\mathcal{G}$ we can define the filter $\mathcal{F}\mathcal{G}$ as that with filter base $\{XY : X \in \mathcal{F}, Y \in \mathcal{G}\}$; similarly, the filters $\mathcal{F}^{-1}$, $\mathcal{F} \vee \mathcal{G}$, $\mathcal{F} \wedge \mathcal{G}$, $con(\mathcal{F})$ and $ocl(\mathcal{F})$ can be defined. For $g \in G$, the principal ultrafilter consisting of all subsets of G which contain g is denoted by $\mathcal{F}(g)$.

A convergence structure is just an assignment $\overset{\star}{\to}$ that tells which filters converge to which ℓ-group elements. Obviously, such an assignment should be well behaved with respect to the ℓ-group structure, and we shall below list some axioms which it should satisfy. We will by no means attempt here to examine the most general such convergence structures possible, but rather consider only enough generality to eventually attain our goal of constructing the lateral completion. The reader should consult Ball's paper for further generality; to make that easier we have identified certain of the axioms below by the names given them by Ball.

An ℓ-convergence structure on an ℓ-group G is an assignment $\overset{\star}{\to}$ which maps filters of G to elements of G satisfying the following:

1. $\mathcal{F}(g) \overset{\star}{\to} g$.
2. If $\mathcal{F} \overset{\star}{\to} g$ and $\mathcal{G} \overset{\star}{\to} g$, then $\mathcal{F} \cap \mathcal{G} \overset{\star}{\to} g$.
3. If $\mathcal{G} \overset{\star}{\to} g$ and $\mathcal{F} \supseteq \mathcal{G}$ then $\mathcal{F} \overset{\star}{\to} g$.
4. The group and lattice operations are continuous.
   That is, if $\mathcal{F} \overset{\star}{\to} f$ and $\mathcal{G} \overset{\star}{\to} k$, then
   (a) $\mathcal{F}^{-1} \overset{\star}{\to} f^{-1}$ and $\mathcal{F}\mathcal{G} \overset{\star}{\to} fk$,
   (b) $\mathcal{F} \wedge \mathcal{G} \overset{\star}{\to} f \wedge k$ and $\mathcal{F} \vee \mathcal{G} \overset{\star}{\to} f \vee k$.
5. ("Hausdorff"): If $\mathcal{F} \overset{\star}{\to} f$ and $\mathcal{F} \overset{\star}{\to} k$, then $f = k$.
6. ("convex"): If $\mathcal{F} \overset{\star}{\to} f$, then $con(\mathcal{F}) \overset{\star}{\to} f$.
7. ("order closed"): If $\mathcal{F} \overset{\star}{\to} f$, then $ocl(\mathcal{F}) \overset{\star}{\to} f$.
8. ("strongly normal"): If $\mathcal{F}^{-1}\mathcal{F}$, $\mathcal{F}\mathcal{F}^{-1}$, and $\mathcal{G} \overset{\star}{\to} 1$ then $\mathcal{F}\mathcal{G}\mathcal{F}^{-1} \overset{\star}{\to} 1$.

The reader may verify that if the notion of convergence arises from a Hausdorff ℓ-topology on the ℓ-group, then all of these properties are satisfied (see Ball [79]). The advantage, which we shall see eventually, is that there also exist ℓ-convergence structures satisfying these axioms which do not arise from a topology.

Notice that the axioms above could easily be restated in terms of filters converging to 1 , just as the axioms for a topological group can be stated in terms of a neighborhood base at the origin (see [Bo], e.g.); we shall leave it to the reader to reformulate these properties in that fashion. Also, since a filter base uniquely determines a filter which then may or may not converge according to a given ℓ-convergence structure, we shall feel free to speak of filter bases converging; it will be understood that we really intend to refer to the corresponding filter.

We can now make sense of Cauchy filters, by saying that $\mathcal{F}$ is Cauchy exactly if $\mathcal{F}\mathcal{F}^{-1}$,

$\mathcal{F}\mathcal{F}^{-1} \xrightarrow{*} 1$. The set $\mathcal{C}$ of Cauchy filters obviously contains the set of convergent ones. Given $\mathcal{F}$, $\mathcal{G} \in \mathcal{C}$, we define the equivalence relation $\sim$ on $\mathcal{C}$ by $\mathcal{F} \sim \mathcal{G}$ iff $\mathcal{F} \cap \mathcal{G} \in \mathcal{C}$. We denote the equivalence class of $\mathcal{F}$ by $[\mathcal{F}]$ and the set of all such by G^x. We then have

Theorem 9.1 G^x is an ℓ-group under the operations

$$[\mathcal{F}][\mathcal{G}] = [\mathcal{F}\mathcal{G}]\ ,\ [\mathcal{F}] \vee [\mathcal{G}] = [\mathcal{F} \vee \mathcal{G}] \text{ and } [\mathcal{F}] \wedge [\mathcal{G}] = [\mathcal{F} \wedge \mathcal{G}].$$

The map $\xi : G \to G^x$ defined by $g\xi = [\mathcal{F}(g)]$ is an ℓ-monomorphism, and $G\xi$ is order dense in G^x.

Proof: Suppose that $\mathcal{F}$ and $\mathcal{G}$ are Cauchy filters. We first must verify that $\mathcal{F}\mathcal{G}$, $\mathcal{F} \vee \mathcal{G}$, and $\mathcal{F} \wedge \mathcal{G}$ are Cauchy. For the first of these, note that

$$\mathcal{H} = (\mathcal{F}\mathcal{G})(\mathcal{F}\mathcal{G})^{-1} = \mathcal{F}\mathcal{G}\mathcal{G}^{-1}\mathcal{F}^{-1};$$

but $\mathcal{G}\mathcal{G}^{-1} \xrightarrow{*} 1$, and so by strong normality $\mathcal{H} \xrightarrow{*} 1$ as required. A similar argument for the other side shows that $\mathcal{F}\mathcal{G}$ is Cauchy. To show that $\mathcal{F} \wedge \mathcal{G}$ is Cauchy, let $\mathcal{L} = \mathcal{F}\mathcal{F}^{-1} \cap \mathcal{G}\mathcal{G}^{-1}$ and $\mathcal{R} = con((\mathcal{L} \vee \mathcal{L}) \cap (\mathcal{L} \wedge \mathcal{L}))$. Clearly, $\mathcal{R} \xrightarrow{*} 1$; we will show that $(\mathcal{F} \wedge \mathcal{G})(\mathcal{F} \wedge \mathcal{G})^{-1} \supseteq \mathcal{R}$ and hence conclude that $(\mathcal{F} \wedge \mathcal{G})(\mathcal{F} \wedge \mathcal{G})^{-1} \xrightarrow{*} 1$. Toward that end, choose $R \in \mathcal{R}$. There exists $L \in \mathcal{L}$ with $con((L \vee L) \cap (L \wedge L)) \subseteq R$, and in turn, there exist $F \in \mathcal{F}$ and $H \in \mathcal{G}$ with $FF^{-1} \cup HH^{-1} \subseteq L$. For $f, k \in F$ and $g, h \in H$,

$$fk^{-1} \wedge gh^{-1} \leq (f \wedge g)(k \wedge h)^{-1} \leq fk^{-1} \vee gh^{-1},$$

and so $(f \wedge g)(k \wedge h)^{-1} \in R$. Hence $(F \wedge H)(F \wedge H)^{-1} \subseteq R$, and so $R \in (\mathcal{F} \wedge \mathcal{G})(\mathcal{F} \wedge \mathcal{G})^{-1}$, as claimed. A similar argument shows that $(\mathcal{F} \wedge \mathcal{G})^{-1}(\mathcal{F} \wedge \mathcal{G}) \xrightarrow{*} 1$, and so $\mathcal{F} \wedge \mathcal{G}$ is Cauchy. We can likewise prove that $\mathcal{F} \vee \mathcal{G}$ is Cauchy if both $\mathcal{F}$ and $\mathcal{G}$ are.

We must now verify that the given operations are well-defined. For that purpose, suppose that $\mathcal{F} \sim \mathcal{F}_1$ and $\mathcal{G} \sim \mathcal{G}_1$.

Then

$$\mathcal{F}\mathcal{G} \cap \mathcal{F}_1\mathcal{G}_1 \supseteq (\mathcal{F} \cap \mathcal{F}_1)(\mathcal{G} \cap \mathcal{G}_1) \in \mathcal{C}$$

and so $\mathcal{F}\mathcal{G} \sim \mathcal{F}_1\mathcal{G}_1$. We leave the arguments for the lattice operations to the reader.

We have thus verified that G^x is an ℓ-group; it is evident that the map ξ is an ℓ-homomorphism. Since $\xrightarrow{*}$ is Hausdorff, it follows that ξ is one-to-one.

Finally, we must show that G is order dense in G^x. For that purpose, suppose that $[\mathcal{F}] \in (G^x)^+$. Now, $[\mathcal{F}] > 1$ means that there exists $F \in \mathcal{F} \vee 1$ with $1 \notin F$, and F is order-closed. Then, we can find $g \in G^+$ with $1 < g \leq F$, or else $1 = \bigwedge F \in F$. //

It is important to observe that if the ℓ-group G belongs to a variety, then so does the ℓ-group G^x constructed above. For if $w = 1$ is an equation not satisfied by G^x, there is a substitution $\bar{w}(h_1, \ldots, h_n)$ of some elements $h_1, \ldots, h_n$ into the variables of w which gives us a nonidentity element of G^x. Let $\mathcal{F}_i$ be a filter converging to h_i, for each i. Then

$\bar{w}(\mathcal{F}_1,\ldots,\mathcal{F}_n)\overset{\star}{\to}h \neq 1.$

But this means we can find $F_i \in \mathcal{F}_i$, with

$1 \notin \bar{w}(F_1,\ldots,F_n).$

That is, we can find $f_i \in F_i$ such that

$\bar{w}(f_1,\ldots,f_n) \neq 1,$

and so the equation $w = 1$ is not satisfied in G either.

Now, the ℓ-group G^x can be viewed as the first step towards an "x-completion" of G. But in order to make sense of this, each ℓ-group must be equipped with a compatible ℓ-convergence structure. Consequently, our next task is to define an assignment of an ℓ-convergence structure for each ℓ-group, in such a way that ℓ-convergence structures on ℓ-groups related by ℓ-homomorphisms are suitably related. Before we do this, we must make some definitions.

Suppose that $\phi : G \to H$ is an ℓ-homomorphism and $\mathcal{F}$ is a filter on G. The homomorphic image of this filter is the filter $\{F\phi : F \in \mathcal{F}\}$, which we denote by $\mathcal{F}\phi$.

Now, suppose that G is an ℓ-subgroup of H, which is equipped with an ℓ-convergence structure $\overset{\star}{\to}$. The restriction of this ℓ-convergence structure to G is denoted by $\overset{\star}{\to}_G$ and means the following: given a filter $\mathcal{F}$ on G, it is a filter base for a filter $\mathcal{G}$ on H. Then $\mathcal{F}\overset{\star}{\to}_G g \in G$ if $\mathcal{G}\overset{\star}{\to}g$.

Next, suppose that X is a subset of G; the *closure of X with respect to the ℓ-convergence structure* $\overset{\star}{\to}$ is

$$\{g \in G : X \in \mathcal{F}\overset{\star}{\to}g\},$$

which is denoted by $cl(X)$. Unfortunately, this closure operator is not necessarily idempotent; we say that a subset of G is *closed* in case $cl(X) = X$. Consequently, we must define the *iterated closure* $itcl(X)$ as the smallest closed subset of G containing X; note that $itcl(X)$ may be obtained by iterating the closure operator (perhaps transfinitely often). If $\mathcal{F}$ is a filter on G, then $cl(\mathcal{F})$ is the filter generated by the filter base consisting of the closures of the elements of $\mathcal{F}$.

We can now make the promised definition. A *convergence* x is an assignment to each ℓ-group of an ℓ-convergence structure $\overset{\star}{\to}$ for that ℓ-group, subject to the following three criteria:

C1. If $\phi : G \to H$ is an ℓ-isomorphism, $\mathcal{F}$ is a filter on G and $\mathcal{F}\overset{\star}{\to}1$, then $\mathcal{F}\phi\overset{\star}{\to}1$.

C2. Suppose that G is a large ℓ-subgroup of H.

(a). Let $\mathcal{F}$ be a filter on G. If $\mathcal{F}\overset{\star}{\to}1$ then $\mathcal{F}\overset{\star}{\to}_G 1$.

(b). If $\mathcal{F}$ is a filter on G and $\mathcal{F}\overset{\star}{\to}1$, then $cl(\mathcal{F})\overset{\star}{\to}1$, where the closure is taken with respect to $\overset{\star}{\to}_G$.

C3. If $h \in G^x$ and $\mathcal{F}$ is a filter on G^x containing G, then $\mathcal{F}\overset{\star}{\to}h$ if and only if $h = [\mathcal{F}]$.

Suppose now that x is a convergence. An ℓ-group H is *x-complete* if $H^x = H$. An ℓ-group H is an *x-completion* of an ℓ-group G if G is large in H, H is x-complete, and if K is any proper ℓ-subgroup of H containing G, then K is not x-complete. Our task now is to inquire into the existence and uniqueness of x-completions.

Theorem 9.2 Suppose that x is a convergence on ℓ-groups. If H is an ℓ-group which is large in H^x, and $\phi : G \to H$ is a one-to-one ℓ-homomorphism such that $G\phi$ is large in H, then there is a unique ℓ-isomorphism $\phi\hat{}$ mapping G^x onto $cl(G\phi)$ in H^x such that $\phi\hat{}$ extends ϕ.

Proof: We define $\phi\hat{}$ by setting $[\mathcal{F}]\phi\hat{} = [\mathcal{F}\phi]$. This is clearly well-defined, because of condition C1; it is obviously an ℓ-homomorphism. Suppose that ψ is another ℓ-homomorphism from G^x into H^x which extends ϕ. Let $[\mathcal{F}] \in G^x$.

Since ϕ and ψ agree on G, $F\phi = F\psi$, for any $F \in \mathcal{F}$. Thus, $[\mathcal{F}]\psi = [\mathcal{F}\psi] = [\mathcal{F}\phi] = [\mathcal{F}]\phi\hat{}$, and so $\phi\hat{}$ is unique.

To show that $\phi\hat{}$ is one-to-one, suppose that $1 \neq [\mathcal{F}] \in G^x$. Since $G\phi$ is large in H, by C1. and C2. $cl(\mathcal{F}\phi)\phi^{-1} \in [\mathcal{F}]$, and so there exists $F \in \mathcal{F}$ with $1 \notin cl(F\phi)\phi^{-1}$. But then, $1 \notin cl(F\phi)$ and so $\mathcal{F}\phi \not\to 1$ in H. Thus, $[\mathcal{F}]\phi\hat{} = [\mathcal{F}\phi] \neq 1$.

It remains to show that $\phi\hat{}$ maps G^x onto $cl(G\phi)$. If $[\mathcal{F}] \in G^x$, then $\mathcal{F} \xrightarrow{\star} [\mathcal{F}]$ and so

$$\mathcal{F}\phi\hat{} = \mathcal{F}\phi \xrightarrow{\star} [\mathcal{F}]\phi\hat{} = [\mathcal{F}\phi] \in H^x,$$

and so $[\mathcal{F}]\phi\hat{} \in cl(G\phi)$. Conversely, if $h \in cl(G\phi)$ in H^x there must be a filter $\mathcal{H}$ on H^x with $G\phi \in \mathcal{H}$ and $\mathcal{H} \xrightarrow{\star} h$. Then by C3, $h = [\mathcal{H}]$. Clearly the filter $\mathcal{H}\phi^{-1}$ is Cauchy in G and so $[\mathcal{H}\phi^{-1}] \in G^x$; thus, $[\mathcal{H}\phi^{-1}]\phi\hat{} = [\mathcal{H}] = h$. Consequently, the image of $\phi\hat{}$ is $cl(G\phi)$. //

We can now determine when x-completions exist:

Theorem 9.3 Suppose x is a convergence on ℓ-groups. Then an ℓ-group G has an x-completion if and only if G is large in some x-complete ℓ-group H. Furthermore, an x-completion of G exists as an ℓ-subgroup of H, namely as $itcl(G)$. The x-completion of G belongs to the same varieties as G.

Proof: The necessity of the condition being obvious, suppose that G is large in an x-complete ℓ-group H. We will show that $itcl(G)$ is an x-completion of G. It is clear that $itcl(G)$ is x-complete, and that G is a large ℓ-subgroup. Now suppose that M is a proper ℓ-subgroup of $itcl(G)$ which contains G. Then $itcl(M)$ properly contains M. We can now apply Theorem 9.2 to the identity map from M into $itcl(M)$ to obtain a lifted map which is an ℓ-isomorphism from M^x onto $itcl(M)$. This show that M is not x-complete, and so $itcl(G)$ is necessarily an x-completion of G . //

We can now characterize x-completions in terms of a map lifting property:

Theorem 9.4 Suppose that x is a convergence on ℓ-groups. Then ℓ-group H is an x-completion of an ℓ-group G if and only if G is large in H, H is x-complete, and every ℓ-monomorphism from G onto a large ℓ-subgroup of an x-complete ℓ-group K can be uniquely lifted to an ℓ-monomorphism from H into K. Consequently, the x-completion of such an ℓ-group is unique up to ℓ-isomorphism over G.

Proof: If H is an ℓ-completion of G, then the previous theorem asserts that $H = itcl(G)$. If we inductively define $G_{\alpha+1} = cl(G_\alpha)$, and $G_\alpha = \bigcup\{G_\beta : \beta < \alpha\}$, for α a limit ordinal, then $H = G_\gamma$, for some ordinal γ. If ϕ is an ℓ-monomorphism from G onto a large ℓ-subgroup of an x-complete ℓ-group K, then Theorem 9.2 guarantees that we may inductively lift ϕ to a map $\phi\hat{}$ from H into K, as claimed. The uniqueness of this map also follows inductively from the uniqueness provision of Theorem 9.2.

Conversely, suppose G is a large ℓ-subgroup of an x-complete ℓ-group H for which the ℓ-monomorphism lifting property holds. We know by Theorem 9.3 that G has an x-completion. If K is any such x-completion, then the ℓ-monomorphism ξ from G into K can be lifted to an ℓ-monomorphism of H into K, which by the usual inductive application of Theorem 9.2 is necessarily onto $itcl_K(G\xi)$, which equals K. Hence, H is ℓ-isomorphic to K, and so is the unique (up to ℓ-isomorphism) x-completion of G. //

We denote the x-completion of a given ℓ-group G by G^{ix}, which stands for iterated x.

We will now apply this machinery to a particular notion of convergence, which will lead us to the existence of a lateral completion, for any ℓ-group. A filter $\mathcal{F}$ on an ℓ-group G *p-converges to 1* if $\bigcap\{F'' : F \in \mathcal{F}\} = 1$, which we shall write as $\mathcal{F}\overset{\star}{\to}1$, as usual; we shall refer to this as *polar convergence*. We now must show that this provides G with an ℓ-convergence structure.

Theorem 9.5 Polar convergence is an ℓ-convergence structure on any ℓ-group.

Proof: We must verify the eight conditions of the definition of an ℓ-convergence structure. Properties 1, 3, 4 and 5 are obvious. Properties 6 and 7 follow immediately, because the double polar of a subset of an ℓ-group is both convex and order closed. To prove property 2, suppose that $\mathcal{F}$ and $\mathcal{K}$ are filters which p-converge to 1 . Now, if $F \in \mathcal{F}$ and $K \in \mathcal{K}$, then $(F \cup K)'' \in \mathcal{F} \cap \mathcal{K}$. But

$$\bigcap(F'' \wedge K'') = \cap F'' \wedge \cap K'' = 1,$$

and so $\mathcal{F} \cap \mathcal{K}\overset{\star}{\to}1$.

Finally, we must show that polar convergence is strongly normal. Suppose that $\mathcal{F}$ and $\mathcal{M}$ are filters such that $\mathcal{F}\mathcal{F}^{-1}$, $\mathcal{M}\overset{\star}{\to}1$. Given $1 < g \in G$, we must show that $g \notin (FMF^{-1})''$, for some $F \in \mathcal{F}$ and $M \in \mathcal{M}$. Toward that end, we choose $F \in \mathcal{F}$ such that $g \notin (FF^{-1})''$; we may as well assume that $g \in (FF^{-1})'$. Choose a specific $f \in F$ and $M \in \mathcal{M}$ such that $f^{-1}gf \notin M''$. Because the set of polars is closed under conjugation, it follows that $g \notin (fMf^{-1})''$. Consequently, we may as well assume that $g \in (fMf^{-1})'$. Since $g \in (FF^{-1})'$, g is disjoint from $|hk^{-1}|$ for all $h, k \in F$, and thus g commutes with any such hk^{-1}. But this means that $h^{-1}gh = k^{-1}kh^{-1}ghk^{-1}k = k^{-1}gk$, for all $h, k \in F$. We can consequently conclude that

$$g \in \bigcap\{(hMh^{-1})' : h \in F\} = (FMF^{-1})';$$

that is, $g \notin (FMF^{-1})''$, as claimed.//

We must next show that polar convergence is in fact a convergence on ℓ-groups:

Theorem 9.6 Polar convergence is a convergence on ℓ-groups.

Proof: We must verify the conditions C1, C2, and C3 of the definition. Condition C1 is obvious, since ℓ-isomorphisms are polar-preserving. Condition C2 follows immediately from the fact that the Boolean algebras of polars of an ℓ-group and a large ℓ-subgroup are isomorphic in the natural way (Theorem 8.1.1). Because polar convergence on G^p reduces to polar convergence on G, to show that condition C3 holds, we need only check that if $\mathcal{F}$ is a Cauchy filter on G, then $\mathcal{F} \overset{\star}{\rightarrow} [\mathcal{F}] = h \in G_p$, or equivalently, that $\mathcal{F}h^{-1} \overset{\star}{\rightarrow} 1$. To do this, we will show that for each $F \in \mathcal{F}$, $Fh^{-1} \subseteq (FF^{-1})''$. Since $\mathcal{F}$ is Cauchy, this will complete the proof. So, suppose that $x \in (FF^{-1})' \cap G$. Now, for any $f \in F$, $|fh^{-1}| \wedge x$ is a filter generated by elements of the form $fK^{-1} \wedge x$, where $K \in \mathcal{F}$. Because $\mathcal{F}$ is a filter, $F \cap K \neq 0$; for k in this intersection, $|fk^{-1}| \wedge x = 1$. Thus, $|fh^{-1}| \wedge x = 1$, which is what we required. //

We now wish to obtain a p-completion for any ℓ-group. By Theorem 9.3, this means that we must show that every ℓ-group is large in a p-complete ℓ-group. We do this by showing that any ℓ-group obtained from an ℓ-group G by iterative application of the p operator has cardinality bounded above by a cardinal dependent only on $|G|$. To do this, we need a definition and a lemma. Given ℓ-groups G and H, we will say that H is a *p-extension* of G if G is (order) dense in H, and for all $h \in H^+$,

$$\bigcap\{(hg^{-1})'' : g \in G\} = 1.$$

Lemma 9.7 Suppose that G and H are ℓ-groups, and H is a p-extension of G. Then $|H| \leq |\{f : \mathcal{P}(G) \rightarrow G\}|$.

Proof: Well-order G and choose a symbol ♣ which does not belong to G. For each $h \in H^+$ we define a function f_h on $\mathcal{P}(H)$ as follows: for $P \in \mathcal{P}(H)$, $f_h(P)$ is the first element g of G such that $hg^{-1} \notin P$, if such exists, and ♣ otherwise. Now suppose that h and k are distinct elements of H^+. Choose a polar P for which $hk^{-1} \notin P$ and $f_h(P) = g \neq$ ♣ (this latter condition can be satisfied because H is a p-extension). Now, if $f_k(P) = g$, then $hk^{-1} = hg^{-1}(kg^{-1})^{-1} \in P$, which is a contradiction. Thus, f_h and f_k are distinct functions. Since G is dense in H, their Boolean algebras of polars are isomorphic. So, because $|H| = |H^+|$ the lemma follows. //

We can now apply this lemma to prove that every ℓ-group has a p-completion:

Theorem 9.8 Every ℓ-group is large in a p-complete ℓ-group; consequently, every ℓ-group admits a unique p-completion.

Proof: Clearly, we need only verify the first statement, from which the second follows, by Theorem 9.3. We inductively define for each ordinal α an ℓ-group G_α by setting $G_0 = G$, $G_{\alpha+1} = G^p_\alpha$, and $G_\alpha = \bigcup\{G_\beta : \beta < \alpha\}$, for limit ordinals α. First, we claim that G_1 is a p-extension of G. We know by Theorem 9.1 that G is order dense in G^p. Now, suppose that $h \in G^{p+}$. Let $\mathcal{F}$ be a filter containing G which converges to h. Let

$\mathcal{M} = \{(Mh^{-1})'' : M \in \mathcal{F}\}$. Since $\mathcal{F}h^{-1} \xrightarrow{*} 1$, it is evident that $\cap\mathcal{M} = 1$. For each $M \in \mathcal{M}$, choose $g \in M \cap G$. Then $(hg^{-1})'' = (gh^{-1})'' \subseteq (Mh^{-1})''$, and so $\bigcap\{(hg^{-1})'' : g \in G\} = 1$, as required. An easy transfinite induction now shows that G_α is a p-extension of G for all ordinals α. But then Lemma 9.7 implies that for some ordinal γ, we must have that $G_\gamma = G_{\gamma+1}$, which is the p-complete ℓ-group in which G is order dense, and consequently large.//

It is not in general the case that the p-completion of an ℓ-group is its lateral completion. However, because the p-completion is laterally complete, the existence of the lateral completion will follow easily.

Theorem 9.10 A p-complete ℓ-group is laterally complete.

Proof: To prove this, we need only show that any pairwise disjoint set in an ℓ-group G has a least upper bound in G^p. For that purpose, suppose that D is a pairwise disjoint subset of G; clearly, we may as well assume that D is in fact an infinite set. For each finite subset E of D, define

$$F(E) = \{\vee L : E \subseteq L \subseteq D, \text{ and } L \text{ is finite }\},$$

and let $\mathcal{F}$ be the filter generated by the sets of the form $F(E)$. We claim that $\mathcal{F}$ is a Cauchy filter. To see this, choose a typical element of $F(E)F(E)^{-1}$; it is of the form $(\vee L)(\vee K)^{-1}$, where L and K are finite subsets of D containing E. Because pairwise disjoint elements commute, it is evident that the double polar of this element is just the double polar of the symmetric difference of the two sets L and K. Consequently, $(F(E)F(E)^{-1})'' \subseteq E' \subseteq D''$, and so the intersection of all such polars is trivial. Thus, we have $[\mathcal{F}] \in G^p$.

It remains to show that $[\mathcal{F}] = \vee D$ in G^p. It is evident that $[\mathcal{F}] \geq d$, for all $d \in D$. Now, suppose that $[\mathcal{F}] \geq h$, some other upper bound for D in G^p. One can check that $[\mathcal{F}](\vee E)^{-1} \in E'$, for all finite subsets E of D (note that the polar operation here is in G^p, but that the Boolean algebras of polars are isomorphic). Since $h \geq \vee E$, this means that

$$[\mathcal{F}]h^{-1} \in \bigcap\{E' : E \text{ a finite subset of } D\} = 1.$$

Hence, $[\mathcal{F}] = h$, as required. //

We can now obtain the result we're after:

Theorem 9.11 Every ℓ-group has a unique lateral completion.

Proof: The previous theorem guarantees that every ℓ-group G can be ℓ-embedded as a dense ℓ-subgroup of a laterally complete ℓ-group, which we will assume is G^{ip}. Since infinite joins are the same in any essential extension of an ℓ-group (Theorem 8.1.2), we certainly obtain a lateral completion G^L by taking the intersection of all laterally complete ℓ-subgroups of G^{ip} containing G.

To prove that this lateral completion is unique, suppose that G has been ℓ-embedded into some other lateral completion H via the map ξ. We may assume that H is a dense

ℓ-subgroup of its p-completion H^{ip}. By the lifting property of theorem 9.4, we obtain an ℓ-monomorphism $\xi^\wedge$ from G^{ip} into H^{ip}. But since ℓ-monomorphisms preserve lateral completeness, $\xi^\wedge(G^L)$ is a laterally complete ℓ-subgroup of H^{ip} which contains (a copy of) G, and consequently contains H, since H is a lateral completion. But then, $(\xi^\wedge)^{-1}(H)$ is a laterally complete ℓ-subgroup of G^L, and consquently equals G^L, because G^L is also a lateral completion. Hence, G^L and H are ℓ-isomorphic over G. //

Notice that it follows immediately that if an ℓ-group belongs to a variety, then its lateral completion does too; this result was first obtained by Bernau [77].

Ball also obtained the Dedekind completion of Theorem 8.2.2 (and its non-archimedean analogue) by considering a different convergence structure; the reader may consult Ball [80a] for details.

Chapter 10: *Finite-valued and Special-valued ℓ-groups*

In this chapter we shall inquire more carefully into the root system of values of an ℓ-group, with the aim of seeing how much information about the ℓ-group itself is contained in this set. The Conrad-Harvey-Holland embedding theorem (Theorem 3.2) serves as a model for this work; we shall discover that many of the results in this chapter, which apply for all normal-valued ℓ-groups, are actually analogues of the abelian case. In particular, we shall determine more precisely the sense in which Read's representation theorem 7.1.9 for normal-valued ℓ-groups is the correct analogue of the Conrad-Harvey-Holland theorem.

Let G be an ℓ-group and $\Gamma(G)$, as usual, its root system of regular subgroups. We recall from Chapter 3 that a subset Δ of $\Gamma(G)$ is plenary if $\cap\Delta = \{1\}$ and Δ is a dual ideal in $\Gamma(G)$. We shall first prove a useful lemma about how a plenary subset of values can be used to determine when an element is positive.

Lemma 10.1 Let G be an ℓ-group and Δ a plenary subset of $\Gamma(G)$. Then $g > 1$ if and only if $Mg > M$, for all values M of g belonging to Δ.

Proof: Since the condition is clearly necessary we now suppose, by way of contradiction, that $g^- \neq 1$. Since $\bigcap \Delta = \{1\}$, $g^- \notin P$, for some $P \in \Delta$. But then g^- has a value $M \supseteq P$, and since Δ is a dual ideal M belongs to Δ. Now M is also a value of g, amd $g^+ \in M$, since $g^+ \wedge g^- = 1$ and M is prime. But then

$$M < Mg = Mg^+(g^-)^{-1} = M(g^-)^{-1} < M,$$

which is a contradiction. //

We are now ready to begin our analysis of $\Gamma(G)$; the first question we shall address is whether there are regular subgroups of G which belong to every plenary subset of $\Gamma(G)$.

To answer this question we make the following definition. A regular subgroup P is an *essential value* if it contains all the values for some $g \in G$. We denote by $E(G)$ the set of essential values of G; since conjugation takes values to values, it is clear that the conjugate of any essential value is essential. We phrase this by saying that $E(G)$ is a normal set of values.

Suppose now that Δ is a plenary subset; let P be an essential value containing all the values of g. Since $\cap\Delta = \{1\}$, this means that $g \notin Q$ for some $Q \in \Delta$. But then g has a value containing Q, which is in turn contained in P, and since Δ is a dual ideal this means that $P \in \Delta$. Thus a plenary subset of $\Gamma(G)$ necessarily contains $E(G)$; we shall enquire later into the question of when $E(G)$ itself is a (necessarily minimal) plenary set.

Note that if Q is the unique value for $g \in G$, then Q is obviously essential; we say in this case that Q is a *special value* for the *special element* g. A suggestive example of this notion is in the Hahn group $V(\Gamma, \mathbf{R})$, where each of the elements of Γ corresponds to the unique value for an element of V living only at that component. Thought about this example leads to the conjecture that if an element of an ℓ-group has only finitely many values, then these values are in fact special. This is in fact the case, and we prove this (and a bit more), in the theorem 10.3 , which is due to Conrad [65]; we first need a technical lemma:

Lemma 10.2 Let $Q_1, Q_2, \ldots, Q_n$ be prime subgroups of an ℓ-group G. Suppose that $1 < x_i \in Q_1 \backslash Q_i$, $1 < y_i \in Q_i \backslash Q_1$, for $i = 1, 2 \ldots, n$. Let $x = \bigvee x_i$ and $y = \bigwedge y_i$. If $h = x(x \wedge y)^{-1}$ and $k = y(x \wedge y)^{-1}$, then

$$h \in Q_1 \backslash \bigcup_{i=2}^{n} Q_i \text{ and } k \in \bigcap_{i=2}^{n} Q_i \backslash Q_1.$$

Proof: Since Q_1 is prime, we clearly have that $x \in Q_1 \backslash \bigcup_{i=2}^n Q_i$ while $y \in \bigcap_{i=2}^n Q_i \backslash Q_1$. But since x and y are incomparable, h and k are strictly positive disjoint elements with $h < x$ and $k < y$. The primeness of the Q_i's then gives the desired conclusion. //

Theorem 10.3 Let G be an ℓ-group, and suppose that $g \in G$ has only finitely many values $Q_1, Q_2, \ldots, Q_n$. Then there exists a representation $g = g_1 g_2 \cdots g_n$, where $|g_i| \wedge |g_j| = 1$, for $i \neq j$, and each g_i is special with unique value Q_i.

Proof: The theorem is trivially true if $n = 1$, and so we shall assume that $n > 1$. Since it is quite clear that the values of g and $|g|$ are the same, it does no harm to assume that g is positive. Furthermore, we may as well assume that $G = G(g)$; this means that the Q_i's are the maximal convex ℓ-subgroups of G; in particular, they are automatically values for elements not belonging to them.

We first show that each Q_i is special. Then Q_1 is a value for k and the other Q_i's are values for h. Now if P is any other value of k, then P is a subset of one of the maximal convex ℓ-subgroups Q_i. But if $i \neq 1$, then $1 = k \wedge h \in G \backslash P$, and so Q_1 is the unique value for k.

We can now pick k_i so that its unique value is Q_i; by replacing k_i by $g \wedge k_i$ if necessary, we may as well assume that $k_i \leq g$. Note that $k_i \wedge k_j = 1$ if $i \neq j$, for if $0 \neq k_i \wedge k_j$ had a value N then both Q_i and Q_j would contain N, which is impossible. But then $\vee G(k_i)$ is a cardinal sum. If g does not belong to this sum then $Q_i \supseteq \Sigma G(k_i)$, for some i; but this is impossible because $k_i \in Q_i$. Thus, $G = G(k_1) \boxplus \ldots \boxplus G(k_n)$ and so

$$g = g_1 \ldots g_n \in G(k_1) \boxplus \ldots \boxplus G(k_n).$$

Now $k_i = g \wedge k_i = g_i \wedge k_i \leq g_i$, and so $G(g_i) = G(k_i)$. This means that each g_i is special with value Q_i, as claimed. //

We are now ready to determine when the set $E(G)$ of essential values forms a plenary subset. To do this we need the following lemma, which describes how certain properties of values can be detected in a plenary subset:

Lemma 10.4 Let G be an ℓ-group, Δ a plenary set of its values, $1 < g \in G$ and $P \in \Gamma(G)$.

a) If g has only finitely many values in Δ then all its values belong to Δ.

b) If all the values of g in Δ are contained in P, then $P \in \Delta$ and all the values of g in $\Gamma(G)$ are contained in P.

Proof: a): Denote the values of g in Δ by $Q_2, \ldots, Q_n$, and suppose by way of contradiction that Q_1 is a value of g which does not belong to the plenary set Δ. By Lemma 10.2, we may obtain strictly positive disjoint elements h, k so that $h \in Q_1 \backslash \bigcup_{i=2}^{n} Q_i$ and $k \in \bigcap_{i=2}^{n} Q_i \backslash Q_1$. Now k has a value $P \in \Delta$, which is contained in some Q_i, for $i = 2, \ldots, n$. But then $1 = h \wedge k \in G \backslash P$, which is a contradiction.

b): Since Δ is a dual ideal it is obvious that $P \in \Delta$. Suppose by way of contradiction that g has a value $Q \in \Gamma(G)$ with $Q \not\subseteq P$. Note that if P itself is a value of g then P is the only value of g in Δ, and hence in $\Gamma(G)$, by part a). Consequently, P is not a value of g and so $g \in P$, since P does contain a value of g.

Since $Q \not\subseteq P$, we may choose $1 < f \in Q \backslash P$. We claim that $gf^{-1} < 1$. To show this, suppose that M is a value of gf^{-1} in Δ. If $M \subseteq P$, then since $f \notin P$ and $g \in P$ we have that $M = P$, and so $g \in M$. If on the other hand $M \not\subseteq P$, then $g \in M$, because all values of g in Δ are contained in P. Thus, in either case we have $Mgf^{-1} = Mf^{-1} < M$. Since this is true for all values of gf^{-1} in the plenary set Δ, this means,by Lemma 10.1, that $gf^{-1} < 1$. But then $1 < g < f \in Q$, which means that $g \in Q$, which is the desired contradiction. //

We can now use this lemma to determine when $E(G)$ is plenary. For suppose that Δ is a plenary subset and $P \in \Delta \backslash E(G)$, and consider $\Lambda = \Delta \backslash \{M : M \supseteq P\}$. We claim that Λ is still plenary: It is obviously a dual ideal. Suppose by way of contradiction that $1 < g \in \cap \Lambda$. Clearly, none of the values of g belong to Λ, and hence all the values of g in Δ are contained in P. But by part b) of Lemma 10.4, all the values of g in $\Gamma(G)$ are contained in P, which makes P essential; this is a contradiction.

Thus, an ℓ-group G has a minimal plenary set of values exactly if $\cap E(G) = \{1\}$, and in this case the minimal plenary subset equals $E(G)$. In general we can define the *Conrad radical* $R(G)$ as the intersection of all the essential values of G; since $E(G)$ is a normal set of values, $R(G)$ is evidently an ℓ-ideal, and of course, $R(G) = \{1\}$ if and only if G has a minimal plenary subset. (Actually, this is not Conrad's original definition [64] of the radical, but is equivalent.) The reader should note that the Conrad radical does not behave as radicals usually do in algebra; namely, for an ℓ-group G, $R(G/R(G))$ need not be trivial (see E34). We record the conclusions of this discussion in the following:

Theorem 10.5 Let G be an ℓ-group. Then the following are equivalent:

a) $E(G)$ is plenary.

b) G has a minimal plenary set of values.

c) $R(G) = \{1\}$.

The reader should again think of the Hahn group $V(\Gamma, \mathbf{R})$; as we have observed in Chapter 3, this ℓ-group has a plenary set of values isomorphic to Γ. In fact, this is a minimal plenary set because, as we mentioned earlier, each of its elements is essential. Of course, for the ℓ-group V each of these essential values is in fact special; we shall generalize this situation later.

Now in order to determine whether a minimal plenary subset exists, we must be able to characterize the essential values. For normal-valued ℓ-groups, the next theorem provides a nice characterization of such values: they are exactly the closed regular subgroups. (Recall the definition of closed convex ℓ-subgroup from Section 1.2. In particular, note that a convex ℓ-subgroup is closed exactly if the natural map to its set of right cosets preserves all meets and joins; this is Theorem 1.2.7.) Byrd [67] proved that an essential value is closed; Byrd and Lloyd [67] proved that the converse holds for normal-valued ℓ-groups.

Theorem 10.6 Every essential value of an ℓ-group is closed; for normal-valued ℓ-groups, every closed value is essential.

Proof: Suppose first that G is a normal-valued ℓ-group, and Q is a closed value for g (where we may as well assume that g is positive); we must show that Q contains all the values of some element of G. To obtain this element, consider the set X of all strictly positive elements of the coset Qg. We claim that X has a strictly positive lower bound. If not then $\wedge X = 1$, and so by Theorem 1.2.7

$$Q = Q(\wedge X) = \bigwedge\{Qx : x \in X\} = Qg,$$

which is a contradiction. We call the lower bound which this contradiction gives us b.

We now claim that all values of b are contained in Q. Suppose instead that P is a value of b and $P \not\subseteq Q$. We can then choose $1 < h \in P\backslash Q$. We can now apply Lemma 4.2.2 to elements g and h to obtain a positive integer n for which $Qh^n > Qg$. But then $Qh^n \wedge g = Qg > Q$, and so $h^n \wedge g \in X$. Thus, $h^n \geq b > 1$, which means that $b \in P$. This contradiction completes the proof that Q is essential.

Conversely, suppose that Q is an essential value. To prove that Q is closed, we suppose by way of contradiction that $\{g_\lambda\}$ is a subset of Q^+, while $g = \vee g_\lambda \notin Q$. We can then find a value P of g such that $P \supseteq Q$.

Since Q is essential, it contains all the values of some $h \in G^+$. Suppose first that $h \in Q$. We shall show that $h^{-1}g \geq g_\lambda$, for all λ, which would be a contradiction. Let N be any value of $h^{-1}gg_\lambda^{-1}$. We claim $h \in N$. If $N \supseteq Q$ this is obvious. Suppose on the other hand that $N \not\supseteq Q$. If $h \notin N$ we can find a value M of h containing N. But then $N \subseteq M \subseteq Q$, which is a contradiction. Thus, we have for all values N of $h^{-1}gg_\lambda^{-1}$ that

$$Nh^{-1}gg_\lambda^{-1} = Ngg_\lambda^{-1} \geq N.$$

But this means, by Lemma 10.1, that $h^{-1}g \geq g_\lambda$, which is the required contradiction.

So, we may now suppose that $h \notin Q$. This means h is special with unique value Q. We claim that $h \wedge g_\lambda = 1$, for all λ. If on the contrary for some λ we have that $h \wedge g_\lambda > 1$, choose a value P of this element. Since $h \notin P$ and Q is the unique value of h, we have that $P \subseteq Q$. But this is contradictory to our assumption, since clearly $h \wedge g_\lambda \in Q$. Hence

$$1 = \vee(h \wedge g_\lambda) = h \wedge (\vee g_\lambda) = h \wedge g.$$

But this contradicts the fact that neither h nor g belongs to the prime Q. This contradiction completes the proof of the theorem. //

Note that this means that an ℓ-group need not have closed regular subgroups (for example E9); thus, although every element of $\mathcal{C}(G)$ is an intersection of regular subgroups, the corresponding statement in the lattice $\mathcal{K}(G)$ need not be true: namely, every closed convex ℓ-subgroup need not be the intersection of closed regular subgroups. We shall mention a result of McCleary's [72b] a bit later which characterizes exactly when this occurs.

Recall from section 1.1 that an ℓ-group is *completely distributive* if and only if it satisfies the *generalized distributive law*: that is,

$$\bigwedge_I \bigvee_J g_{ij} = \bigvee_{f \in J^I} \bigwedge_I g_{if(i)},$$

whenever the indicated meets and joins exist (of course, this law always holds if the index sets I and J are finite). As we shall see, the notion of complete distributivity is closely related to the idea of a minimal plenary subset, at least in the normal-valued case.

To describe this connection, we need the following definition: Given an ℓ-group G, let $D(G)$ be the intersection of all closed prime subgroups of G; $D(G)$ is called the *distributive radical*. We can make use of this radical to characterize completely distributive ℓ-groups in the following theorem, due to Byrd and Lloyd [67]. Note that this theorem can be proved by elementary ℓ-group means; for that proof the reader should consult Byrd and Lloyd [67] or Bigard, Keimel and Wolfenstein [BKW]. An important role in that proof is played by Weinberg's characterization [62] of complete distributivity in terms of "distributive pairs". We shall not describe this work here, but instead reduce the theorem to considerations about complete distributivity in $A(T)$, for which the reader may refer to Glass's book [G].

Theorem 10.7 An ℓ-group G is completely distributive if and only if $D(G) = \{1\}$.

Proof: Suppose first that $D(G) = \{1\}$. We can then define an ℓ-embedding from G into the ℓ-group $A(T)$ like that used in the proof of Holland's Theorem 5.4, where T is a totally ordered set constructed lexicographically from the totally ordered sets of right cosets of the form G/P, where P is a closed prime. Because the P's are closed, this map preserves all meets and joins (Theorem 1.2.7). It can then be verified (see [G] Theorem 8.2.5) that all suprema and infima in the ℓ-subgroup of $A(T)$ ℓ-isomorphic to G can be computed in $A(T)$ using pointwise suprema and infima. But then the complete distributivity of G is reduced to the evident complete distributivity of the chain T.

Conversely, suppose that G is completely distributive, and let $g \in D(G)^+$. Let $\{P_\alpha : \alpha \in A\}$ denote the set of all primes of G. Now for each α, $g \in \bar{P}_\alpha$, the closure of P_α. But then $g = \bigvee_B g_{\alpha\beta}$, for $\{g_{\alpha\beta} : \beta \in B\} \subseteq P_\alpha$. We thus have

$$g = \bigwedge_A \bigvee_B g_{\alpha\beta} = \bigvee_{f \in B^A} \bigwedge_A g_{\alpha f(\alpha)}.$$

But the last meet belongs to all primes of G, and is consequently 1. Thus $g = 1$, and so $D(G) = \{1\}$, as required. //

An important and difficult related result first proved by Lloyd [64] is that the ℓ-group $A(T)$ is always completely distributive; this follows from the fact that the point stabilizers are closed primes.

Note that McCleary [72b] has proved (using permutation group methods) that complete distributivity is equivalent to every closed convex ℓ-subgroup being an intersection of closed primes, contrary to the usual situation discussed after the proof of Theorem 10.6; for this proof see [G].

For normal-valued ℓ-groups complete distributivity reduces to a situation with which we are already familiar:

Corollary 10.8 For a normal-valued ℓ-group G, $D(G) = R(G)$, and so G is completely distributive if and only if $R(G) = \{1\}$; that is, if and only if G has a minimal plenary subset.

Proof: Since every essential value is closed, it is obvious that $D(G) \subseteq R(G)$. The opposite inclusion follows, because each closed prime is obviously contained in a value, and the following lemma asserts that a prime containing a closed prime is closed. //

To conclude the proof of the last corollary, we must prove the following lemma, due to Byrd and Lloyd [67]:

Lemma 10.9 A prime containing a closed prime is closed.

Proof: Suppose that P is a closed prime and Q is a prime which properly contains it; choose $h \in Q^+ \backslash P$. Suppose further that $\{g_\lambda\} \subseteq Q^+$ and $g = \vee g_\lambda$; we will show that $g \in Q$. Since P is closed we have by Theorem 1.2.7 that $Pg = \vee(Pg_\lambda)$. Now, if $g \wedge h \in P$, we have that $g \in P \subseteq Q$, since P is prime. So, we shall suppose that $g \wedge h \notin P$. This means that

$$P \leq P(g \wedge h)^{-1}g < Pg.$$

We thus have that $P(g \wedge h)^{-1}g \leq Pg_\lambda$, for some λ. But then there exists $p \in P$ such that

$$1 \leq (g \wedge h)^{-1}g \leq pg_\lambda \in Q.$$

Thus $(g \wedge h)^{-1}g \in Q$, and so $g \in Q$. //

Note that a particular consequence of Corollary 10.8 is that the ℓ-group $V(\Gamma, \mathbf{R})$ is completely distributive; thus the target ℓ-group of both the Holland and the Conrad-Harvey-Holland representation theorems are completely distributive. However, Conrad [64] has shown that for archimedean ℓ-groups complete distributivity is equivalent to having a basis and thus the target ℓ-group $D(X)$ for the Bernau representation theorem is almost never completely distributive; we will not explicate this work here.

We shall now examine two cases where a normal-valued ℓ-group has a particularly nice minimal plenary subset. We say that an ℓ-group is *finite-valued* if every element has only finitely many values; by Theorem 10.3 this obviously implies that every value is special; the next theorem asserts that the converse holds as well. Note that since special values are essential, this is clearly a special case of the circumstance where there are no proper plenary subsets. However, there exist ℓ-groups with no proper plenary subsets which are not finite-valued; see E38.

The appropriate abelian example of a finite-valued ℓ-group to think of here is the ℓ-subgroup $\Sigma(\Gamma, \mathbf{R})$ consisting of those elements of the Hahn group with finite support. It is obvious in this case that every value is special.

Theorem 10.10 For an ℓ-group G the following are equivalent:

a) G is finite-valued.

b) Every element of G can be expressed as a product of finitely many pairwise disjoint special elements.

c) Every value is special.

Proof: The equivalence of a) and b) follows easily from Theorem 10.3, and clearly b) implies c).

Suppose then that every value of the ℓ-group G is special. Suppose by way of contradiction that $g \in G$ has infinitely many values $\{Q_\lambda\}$. Choose for each Q_λ an element h_λ whose unique value is Q_λ. We then know that the h_λ's are pairwise disjoint, and so the join of the convex ℓ-subgroups $G(h_\lambda)$'s is their cardinal sum. Now $g \notin \Sigma G(h_\lambda)$, and so g has a value M which contains $\Sigma G(h_\lambda)$; but this is impossible, because M must be one of the Q_λ's . //

An important observation to make about finite-valued ℓ-groups is that they are normal-valued (and hence completely distributive, by Corollary 10.8):

Proposition 10.11 A special value is normal in its cover. Hence, a finite-valued ℓ-group is normal-valued.

Proof: Suppose an ℓ-group G has a special value P not normal in its cover. Then by Lemma 4.2.2 there exist $x, y \in G \setminus P$ such that $Px^n \le Py$, for all integers n. We may assume without loss of generality that P is a maximal convex ℓ-subgroup of G and so $P^* = G$. Now y can be written as a join of finitely many pairwise disjoint special elements (by Theorem 10.10), and P is the value of exactly one of these, which determines the same coset as y. Thus we may as well assume that y is special.

Now consider the conjugate $x^{-1}Px$, a maximal convex ℓ-subgroup of G. We claim that $y \notin x^{-1}Px$; we thus suppose the contrary. Then $y = x^{-1}qx$, for some $q \in P$. Since $Px > P$, we have that $px > 1$, for some $p \in P$. But then $y = x^{-1}qx < pqx$, and so $Py \le Px < Py$, which is a contradiction.

In this case $x^{-1}Px$ (being a maximal convex ℓ-subgroup) is a value for y, and since y is special this means that $P = x^{-1}Px$. Now $G = P \vee G(x)$, and so $y \le \prod_{i=1}^n (p_i x)$, for $p_i \in P$. Hence $y \le (rx)^n$, where $r = \vee p_i \in P$. But then $y \le (rx)^n \le sx^n$ for some $s \in P^+$, since $xP = Px$. This means that $Py \leq Px^n$, which contradicts the fact that $Px^n < Py$. This contradiction proves the theorem. //

We are now able to show that an ℓ-group is finite-valued exactly if its lattice of convex ℓ-subgroups is *freely generated* by its root system of regular subgroups; by this we mean that there is a one-one correspondence between elements of $\mathcal{C}(G)$, and dual ideals of $\Gamma(G)$; given a convex ℓ-subgroup C, the dual ideal is obtained by looking at all regular subgroups containing it. This will obviously always give us a dual ideal whose intersection is C; the difference in this circumstance is that the dual ideal whose intersection is C is unique. This is what must be proved in order to verify the following theorem, which is due to Conrad [65]:

Theorem 10.12 An ℓ-group is finite-valued if and only if its lattice of convex ℓ-subgroups is freely generated by its root system of regular subgroups.

Proof: Suppose that G is finite-valued and that $C \in \mathcal{C}(G)$. Let Λ be a dual ideal of regular subgroups such that $\cap\Lambda = C$; we must show that Λ is the dual ideal of *all* regular subgroups which contain C. Thus, suppose that P is regular and $P \supseteq C$. But P is the unique value of a special element g. Since $g \notin C = \cap\Lambda$, there exists $Q \in \Lambda$ such that $g \notin Q$. But then $P \supseteq Q$, and so $P \in \Lambda$.

Conversely, suppose that the lattice of convex ℓ-subgroups is freely generated by the root system of regular subgroups. Given a regular subgroup P, we must show that P is special. But consider the distinct dual ideals of $\Gamma(G)$

$$\{Q : Q \not\subseteq P\} \text{ and } \{Q : Q \not\subset P\};$$

these dual ideals must have distinct intersections, and consequently we can find

$$1 < g \in \bigcap\{Q : Q \not\subseteq P\} \backslash \bigcap\{Q : Q \not\subset P\}.$$

It is easy to check that P is the unique value of g. //

Taking our inspiration from the Hahn group $V(\Gamma, \mathbf{R})$, we shall now generalize the notion of finite-valued a bit: namely, we shall consider ℓ-groups which have a plenary subset of special values (which is then necessarily the minimal plenary set of all essential values). We call such ℓ-groups *special-valued*. We first prove that special-valued ℓ-groups are normal-valued (and hence completely distributive); this is a corollary of the following theorem (first proved by Wolfenstein in his thesis [70a]; for a permutational proof, see [G]):

Theorem 10.11 An ℓ-group with a plenary set of normal values is normal-valued.

Proof: Suppose that G has a plenary set Δ of normal values. We shall check that G satisfies the inequality (c) in Theorem 4.2.4 which characterizes normal-valued ℓ-groups, that is, written multiplicatively, $ab \leq b^2a^2$, for all $a, b \in G^+$.

So suppose that $a, b \in G^+$; by Lemma 10.1, we need only check that $Pab \leq Pb^2a^2$ for all $P \in \Delta$. Thus we shall assume that for some P, $Pab > Pb^2a^2$. But because Δ is plenary, we may as well suppose that P is a value for ab. But then $Pa, Pb \in P^*/P$, an abelian o-group, and so we have $Pab > P(ab)^2$, which is absurd.//

Corollary 10.14 A special-vaues ℓ-group is normal-valued.

Proof: This follows immediately from Proposition 10.11. //

We can now prove the analogue of Theorem 10.10, which Conrad proved in [67]:

Theorem 10.15 An ℓ-group is special-valued if and only if each positive element can be expressed as the join of a set of pairwise disjoint positive special elements.

Proof: Suppose first that G is an ℓ-group in which each positive element can be expressed as a join of pairwise disjoint special elements. Let Δ be the set of special values; we must show that Δ is plenary.

Given $1 < g \in G$, we have that $g = \vee g_\lambda$, where $\{g_\lambda\}$ is a set of positive special elements. Let $P \in \Delta$ be the value of g_λ; then P is also a value of g and so $g \notin P$. Thus, $\cap\Delta = \{1\}$.

To show that Δ is also a dual ideal, suppose that $P \in \Delta$ and Q is a regular subgroup with $Q \supset P$. Now Q is a value for some $1 < g = \vee g_\lambda$, where $\{g_\lambda\}$ are special. Because P is special it is closed (Theorem 10.6) and so Q is also closed (Lemma 10.9). But then $g_\lambda \notin Q$, for some λ, and so its unique value R contains Q. Since R is also a value for g, this means that $R = Q$, and so Q is special, as required.

Conversely, suppose that the set Δ of special values is plenary. Let $1 < g \in G$. For each special value P_λ of g, choose a positive special element h_λ whose value is P_λ. Since P_λ^*/P_λ is an archimedean o-group, we may as well assume that $P_\lambda h_\lambda > P_\lambda g$ (replacing h_λ by h_λ^n as necessary). Now let $g_\lambda = h_\lambda \wedge g$; we claim that g_λ is special. To show this, consider any $Q \in \Delta$. If $Q \supseteq P_\lambda$, then clearly $Qh_\lambda > Qg$, and so $Qg_\lambda = Qg$. If $Q \not\supseteq P_\lambda$, then $h_\lambda \in Q$, and so

$$Qg = (Qg) \wedge (Qh_\lambda) = Q.$$

Thus Q is the only value of g_λ in Δ and hence in $\Gamma(G)$, by Lemma 10.4.

It remains to show that $g = \vee g_\lambda$. For that purpose, suppose that $h \geq g_\lambda$, for all λ; we must show that $hg^{-1} \leq 1$. Let M be a value of hg^{-1} in Δ. If $g \in M$, then $Mgh^{-1} = Mh^{-1} \leq M$. If $g \notin M$, then $P_\lambda \supseteq M$ for some λ, and so $Mgh^{-1} = Mg_\lambda h^{-1} \leq M$. Thus, $gh^{-1} \leq 1$ as required. //

The next theorem provides the analogue for special-valued ℓ-groups of Theorem 10.12. Namely, it asserts that an ℓ-group is special-valued exactly if there is a one-one correspondence between its lattice of closed convex ℓ-subgroups and the dual ideals of its root system of closed regular subgroups (that is, the closed regular subgroups freely generate the lattice of closed convex ℓ-subgroups). Conrad [81] proved that the condition is sufficient; Bixler and Darnel [86] proved the converse. Since a special-valued ℓ-group has a minimal plenary subset and so is completely distributive, this theorem can be fruitfully compared to McCleary's result [72b] mentioned earlier that an ℓ-group is completely distributive if and only if each closed convex ℓ-subgroup is the intersection of the closed primes containing it.

Theorem 10.16 An ℓ-group is special-valued if and only if its lattice of closed convex ℓ-subgroups is freely generated by its root system of closed regular subgroups.

Proof: First suppose that G is a special-valued ℓ-group. We shall first prove that each closed convex ℓ-subgroup is in fact the intersection of the closed regular subgroups which contain it; of course, this actually follows from McCleary's general result about completely distributive ℓ-groups, but for special-valued ℓ-groups we can provide an easy proof. Namely, suppose that K is a closed convex ℓ-subgroup of G, and let $g \in G^+ \backslash K$. Because G is special-valued, $g = \vee g_\lambda$, where each g_λ is a (positive) special element. Because K is closed $g_\lambda \notin K$, for some λ. But then g_λ has a value which contains K; since g_λ is special this value is unique and hence closed. Thus K is the intersection of a set of closed regular subgroups.

It thus follows that every closed convex ℓ-subgroup is the intersection of the dual ideal of all closed regular subgroups which contain it; we must now show that this dual ideal is unique. So suppose that Λ is a dual ideal of closed regular subgroups whose intersection is the closed convex ℓ-subgroup K, but that P is a closed regular subgroup containing K with $P \notin \Lambda$. Since P is closed it is essential (by Theorem 10.6), and consequently is special, since G is special-valued. Let g be a (positive) special element with unique value P. Now consider $Q \in \Lambda$. If $g \notin Q$, then $P \supset Q$ and so $P \in \Lambda$, because Λ is a dual ideal; thus $g \in Q$ for all $Q \in \Lambda$. This means that $g \in \cap\Lambda = K$, which is a contradiction.

Conversely, suppose that the lattice of closed convex ℓ-subgroups of the ℓ-group G is freely generated by the closed regular subgroups of G. By Lemma 10.9 the set of closed regular subgroups is a dual ideal of $\Gamma(G)$, and because $\{1\}$ is a closed convex ℓ-subgroup it is the intersection of the closed regular subgroups; consequently they form a plenary subset of $\Gamma(G)$.

It thus remains to show that each closed regular subgroup P is special; we do this by means of the same argument as for Theorem 10.12: The sets of closed regular subgroups

$$\{Q : Q \nsubseteq P\} \quad \text{and} \quad \{Q : Q \not\subset P\}$$

are distinct dual ideals and so have distinct intersections. We can thus find

$$1 < g \in \bigcap\{Q : Q \nsubseteq P\} \backslash \bigcap\{Q : Q \not\subset P\},$$

which is clearly special with value P. //

Obviously the Hahn group $V(\Gamma, \mathbf{R})$ is special-valued; the assertion of the theorem about the closed convex ℓ-subgroups of V is a theorem of Anderson and Conrad's [81].

Now the Conrad-Harvey-Holland embedding theorem (Theorem 3.2) asserts that every abelian ℓ-group can be ℓ-embedded into a special-valued (abelian) ℓ-group. We can now examine Read's representation theorem for normal-valued ℓ-groups (Theorem 7.1.9) in this light: it asserts that every normal-valued ℓ-group can be ℓ-embedded into a special-valued ℓ-group (which is necessarily normal-valued). In order to see this, we need to look more closely at Read's representation.

Read's theorem asserts that every normal-valued ℓ-group can be ℓ-embedded into a generalized wreath product $W = \mathrm{Wr}\{\mathbf{R}_\gamma : \gamma \in \Gamma\}$ of copies of the reals indexed by the totally ordered set Γ (to make the discussion which follows easier, we will use multiplicative notation, and consequently will understand by $\mathbf{R}$ the multiplicative o-group of positive reals). Recall from Section 7.1 that elements of W can be expressed as certain $\Gamma \times V$ "matrices", where V is the usual abelian Hahn group $V(\Gamma, \mathbf{R})$; these elements can then be viewed as permutations of the totally ordered set V, where a given element of V is multiplied by the corresponding column in the matrix.

We first wish to observe that this representation is indeed nonoverlapping, in the sense described in Chapter 5, and consequently the version of Read's theorem mentioned there is implied by the wreath product representation of Theorem 7.1.9. To see this, we must examine the convex subset $\Delta(v, g)$ of V swept out by images of a given (positive) element g of W. Consider the maximal element in the support of the vth column of the matrix for g and the maximal element in the support of v; let γ be the larger of these two elements of Γ. The images of v by powers of g will then all have maximal component at γ, and since the coordinates here are just real numbers, we obtain for $\Delta(v, g)$ the set of all elements of V whose maximal component is γ. Any two such sets are comparable, and so we have that W has a nonoverlapping representation, as claimed.

We now wish to show that the wreath product W is in fact a special-valued ℓ-group. To do this, we shall define elements of W as follows, which we shall show are special. Let $1 < b \in \mathbf{R}$. Choose $\gamma \in \Gamma$ and let T be one of the K^γ-equivalence classes of V. Of course, T is a convex subset of V; choose $t \in T$. Consider then the element $b_{\gamma,T}$ of W defined by the matrix which is b at all entries in the γth row with column coordinates from T, and 1 elsewhere. Now let P consist of all the elements of W which are 1 at all locations δ, v, where $\delta \geq \gamma$ and $vK^\gamma t$. It is evident that P is a value of $b_{\gamma,T}$. Suppose now that N is another value of $b_{\gamma,T}$, and so we can find $g \in P^+ \backslash N$; but then some power of g would exceed $b_{\gamma,T}$, which is absurd. Consequently, P is a special value.

Now, any convex ℓ-subgroup of W contains an element of the form $b_{\gamma,T}$, for some γ and some T, and so the intersection of these special values is obviously $\{1\}$. It is also quite evident that the set of these values is a dual ideal, and consequently W has a plenary set of special values, and so is special-valued.

We thus have the following corollary to Read's Theorem 7.1.9:

Corollary 10.17 An ℓ-group is normal-valued if and only if it can be ℓ-embedded into a special-valued ℓ-group.

An even closer analogy to the Conrad-Harvey-Holland theorem is possible, if we interpret it in its stronger formulation. We recall that it actually asserts that any (abelian) ℓ-group with a plenary set Δ of values can be ℓ-embedded into an (abelian) special-valued ℓ-group, whose minimal plenary set of special values is naturally isomorphic to Δ. Ball, Conrad and Darnel [86] have proved that this is the case if "abelian" is replaced by "normal-valued" (when the plenary set of values is normal). Note that Ball, Conrad and Darnel failed to observe that the plenary set of values used need be normal; Bixler and Darnel [86] corrected this. The original proof makes use of Ball's theory of Cauchy structures from Chapter 9; Bixler and Darnel have since provided a proof by elementary ℓ-group means. However, we shall present the original proof, since it bears the advantage of allowing us to conclude that the target of the embedding belongs to the same variety as the original ℓ-group (which is of course true for the Conrad-Harvey-Holland Theorem); we need this in order to prove the theorem on the nonrepresentable covers of the abelian variety, as promised in Section 7.2. As we did for the Conrad-Harvey-Holland theorem, we shall actually prove the Ball-Conrad-Darnel theorem only for the full set of all values, and leave the modification necessary for a normal plenary subset to the reader.

Theorem 10.18 Each normal-valued ℓ-group G can be ℓ-embedded into a special-valued ℓ-group H in such a way that if Δ is the plenary set of special values for H, then

$$\Gamma(G) = \{G \cap P : P \in \Delta\}.$$

Furthermore, H belongs to the same ℓ-group varieties as G.

Proof: Let G be a normal-valued ℓ-group, and let

$$\mathcal{C} = \{Y \subseteq G : Y \supseteq \cap F, F \text{ a finite subset of } \Gamma(G)\}.$$

This provides a neighborhood filter for a topology on G making the lattice and group operations continuous, and this induces an ℓ-convergence structure on G whose Cauchy filters are exactly

$$\mathcal{D} = \{\text{filters } \mathcal{F} : (\mathcal{F}\mathcal{F}^{-1}) \cap (\mathcal{F}^{-1}\mathcal{F}) \supseteq \mathcal{C}\}.$$

We now consider G^x, the ℓ-group constructed in Theorem 9.1. For each $P \in \Gamma(G)$, consider

$$cl(P) = \{[\mathcal{F}] \in G^x : P \in \mathcal{F}\},$$

the closure of P in G^x with respect to our convergence structure. It is quite evident that the set of such is a collection of regular subgroups of G^x, which we denote by Δ. The density of G in G^x implies that $\cap\Delta = \{1\}$, and so to show that Δ is a plenary subset of $\Gamma(G^x)$ we need only check that it is a dual ideal.

For that purpose, choose $P \in \Gamma(G)$ and $Q \in \Gamma(G^x)$ with $Q \supseteq cl(P)$. Let $[\mathcal{F}] \in G^x$ have value Q. Then there exists $F \in \mathcal{F}$ such that $F \cap Q \cap G = \{1\}$, or else $[\mathcal{F}] \in cl(Q \cap G) \subseteq Q$. But then $f \in F$ has value $R \in \Gamma(G)$ such that $R \supseteq Q \cap G \supseteq P$, and clearly $Q = cl(R)$.

We now must show that G^x is special-valued, where Δ is its plenary set of special values. Let $h = [\mathcal{H}] \in G^x$; then h has a value $cl(P) \in \Delta$, since Δ is plenary. We will now construct a component of h whose unique value is $cl(P)$; this will then show that G^x is special-valued, by Theorem 10.15. To do this, let

$$\mathcal{M} = \{Qg \in \mathcal{H} : Q \in \Gamma(G) \text{ and } Q \subseteq P\}$$

and

$$\mathcal{N} = \{Q \in \Gamma(G) : Q \text{ and } P \text{ are incomparable}\}.$$

We claim that every finite subset of $\mathcal{F} = \mathcal{M} \cup \mathcal{N}$ has nonempty intersection and so serves as a filter base. For this purpose, suppose that $\mathcal{M}_1$ and $\mathcal{N}_1$ are finite subsets of $\mathcal{M}$ and $\mathcal{N}$, respectively. For each $M_i \in \mathcal{M}_1$, choose $g_i \in M_i \backslash P$. Then

$$g = \wedge g_i \in \cap M_i \backslash P.$$

Since $\mathcal{M} \subseteq \mathcal{H}$, we may choose $h \in \mathcal{M}_1$. Because G is normal-valued, there exists (by Lemma 4.2.2) a positive integer n such that $Ph < Pg^n$. Thus,

$$g^n \in h \in (\bigcap \mathcal{M}_1) \cap (\bigcap \mathcal{N}_1),$$

and so the intersection is nonempty, as claimed.

Thus $\mathcal{F}$ serves as a filter base for a Cauchy filter; let $[\mathcal{F}] = f \in G^+$. We must show that f is a component of h. Suppose that $cl(Q) \in \Delta$. If $Q \subseteq P$, then there exists $g \in G$ such that $Qg \in \mathcal{H} \cap \mathcal{F}$, and so $Q \in \mathcal{H}\mathcal{F}^{-1}$; on the other hand, if Q and P are unrelated, then $Q \in \mathcal{F}$. Therefore, $(\mathcal{H}\mathcal{F}^{-1}) \wedge \mathcal{F} \supseteq \mathcal{C}$, and so $(hf^{-1}) \wedge f = 1$; that is, f is a component of h.

Now, $f \in cl(Q)$, for all $cl(Q) \in \Delta$ such that $cl(Q)$ is unrelated to P, while $hf^{-1} \in cl(R)$, for all $cl(R) \in \Delta$ such that $P \supset R$. This means that f is special with unique value $cl(P)$, as required.

The fact that the ℓ-group H belongs to the same varieties as G follows immediately from Theorem 9.3. //

Of course, the target of the Conrad-Harvey-Holland theorem is laterally complete, in addition to being special-valued. This is actually also the case here (see Ball, Conrad and Darnel [86] and also Bixler and Darnel [86]).

Corollary 10.19 Every normal-valued variety is determined by its special-valued ℓ-groups.

We shall now present Darnel's [86c] proof of the fact (due first to Gurchenkov and Kopytov [87] and subsequently and independently to Reilly [86] and Darnel [86c]) that the Scrimger varieties $\mathcal{S}_p$ are the only nonrepresentable covers of the abelian variety $\mathcal{A}$ in the lattice $\mathbf{V}$ of ℓ-group varieties. We need a few preliminaries first.

We should recall a bit about the Scrimger varieties from Section 7.2. For each prime p the Scrimger variety $\mathcal{S}_p$ is the variety generated by an ℓ-subgroup G_p of $\mathbf{Z}$ Wr $\mathbf{Z}$, and these varieties are nonrepresentable covers of $\mathcal{A}$. Recall that Smith's Lemma 7.2.5 provides a characterization of the ℓ-groups G_n in terms of relations on generators.

Next we need to make an observation about values of disjoint elements. Suppose that a and b are pairwise disjoint elements of an ℓ-group G. Then it clearly follows that they have no values in common and no value of a is comparable to a value of b, because values are prime. In fact, the converse of this is true: if a and b are positive elements with no values in common, and no comparable values, then $a \wedge b = 1$. For if not, choose a value P of $a \wedge b$; but then a and b both have values containing P, and since P is prime these values must be comparable (Theorem 1.2.10).

Finally, we need three lemmas:

Lemma 10.20 The normalizer of a convex ℓ-subgroup of an ℓ-group is an ℓ-subgroup.

Proof: Let C be a convex ℓ-subgroup of an ℓ-group G, and let N be the normalizer of C. Given $a, b \in N$, we need to check that $a \vee b \in N$. Let $g \in C^+$. Then

$$1 \le (a \vee b)^{-1} g (a \vee b) = (a^{-1} \wedge b^{-1}) g (a \vee b)$$
$$= (a^{-1} ga \wedge b^{-1} ga) \vee (b^{-1} ga \wedge b^{-1} gb) \le a^{-1} ga \vee b^{-1} gb \in C,$$

and so $a \vee b \in N$, as required.//

Lemma 10.21 Let G be a normal-valued ℓ-group and $c \in G^+$. Then $\{g \in G : |g| \ll c\}$ is a convex ℓ-subgroup of G.

Proof: Let H be the subset of G given in the statement of the lemma. It is obvious that H is convex; to complete the proof we need only verify that $g \in H$ if and only if g belongs to all values of c intersected with $G(c)$.

Let P be a value of c and suppose that $g \in G^+ \backslash P$, while $g \ll c$. Now $g, c \in P^*$, and so $Pg^n \le Pc$ for all n, but this contradicts the fact that P^*/P is archimedean.

Conversely, suppose that $g \in G(c)$, belongs to all values of c and $g^n \not\le c$, for some n. Following the proof of (a) $\Rightarrow$ (b) in Theorem 4.2.4, we can find a value P of c for which $Pg^n > Pc$. Hence $g \notin P$, which is a contradiction. //

Lemma 10.22 Let G be a normal-valued ℓ-group, and suppose that $a, b, g \in G$ with $1 < a, g \ll b^{-1} ab$ and $a \wedge g^{-1} ag = 1$. Then G contains a copy of $\mathbf{Z}$ wr $\mathbf{Z}$.

Proof: First note that $1 < a \ll b^{-1} ab$ implies that $1 < a \ll b$. For assume that $a^n \not\le b$, for some n. If $a^n \ge b$, then $b^{-1} ab \le aa^n$. So we may assume a^n and b are incomparable. Let $d = b \wedge a^n$ and $c = a^n - b \wedge a^n$. Then $a^n = c + d$, $c \wedge d = c \wedge b = 1$, and $c \ne 1$. Hence we can find a value P of c with $b \in P$; P is then also a value of a^n. Now $a \notin P$, therefore $b^{-1} ab \notin b^{-1} Pb = P$. $a \not\ll b^{-1} ab$ follows from the fact that P^*/P is archimedean.

So, we need now only check that $g^{-1} ag$ has infinitely many pairwise disjoint conjugates by powers of b, which lie infinitely below b.

Since $|g|, a \ll b^{-1} ab$ it follows from Lemma 10.21 that $g^{-1} ag \ll b^{-1} ab$ and hence $g^{-1} ag \ll b^{-n} ab^n$ for all positive integers n. Then for any positive integers n and m with $1 \le m < n$ we have

$$1 \le b^{-n} g^{-1} agb^n \wedge b^{-m} g^{-1} agb^m \le b^{-n} g^{-1} agb^n \wedge b^{-m} (b^{m-n} ab^{n-m}) b^m$$
$$= b^{-n} g^{-1} agb^n \wedge b^{-n} ab^n = b^{-n} (g^{-1} ag \wedge a) b^n = 1. \qquad //$$

Theorem 10.23 The only nonrepresentable covers of $\mathcal{A}$ are the Scrimger varieties.

Proof: It suffices to show that if $\mathcal{V} \supset \mathcal{A}$ is a variety of ℓ-groups that contains none of the Scrimger ℓ-groups, then $\mathcal{V}$ must contain a nonabelian o-group, because any cover of $\mathcal{A}$ that contains a nonabelian o-group must be contained in $\mathcal{R}$.

Clearly $\mathcal{V} \subseteq \mathcal{N}$ since $\mathcal{V}$ contains no Scrimger ℓ-groups and $\mathcal{N}$ is the largest proper variety. Assume that G is a nonrepresentable ℓ-group in $\mathcal{V}$. By Corollary 10.16, we may assume that G is special-valued.

Since G is nonrepresentable, there exist $a, b \in G^+$ with $b^{-1}ab \wedge a = 1$, by Theorem 4.1.1. Since G is special-valued, each positive element is the join of disjoint special elements. Hence, we may assume that a (and hence $b^{-1}ab$) is special-valued. For if $a = \vee a_i$, each a_i special, then

$$1 = b^{-1}ab \wedge a = b^{-1}(\vee a_i)b \wedge (\vee a_i) = \vee(b^{-1}a_i b \wedge a_i).$$

Hence $b^{-1}a_i b \wedge a_i = 1$ for all a_i.

If for all $n > 0$, $b^{-n}ab^n \wedge a = 1$, then the ℓ-group G generated by a and b is ℓ-isomorphic to $\mathbf{Z}$ wr $\mathbf{Z}$, which leads to a contradiction since if $\mathbf{Z}$ wr $\mathbf{Z}$ is in $\mathcal{V}$ then its lateral completion $\mathbf{Z}$ Wr $\mathbf{Z}$ is in $\mathcal{V}$, by the remarks following Theorem 9.11, and each Scrimger ℓ-group is an ℓ-subgroup of $\mathbf{Z}$ Wr $\mathbf{Z}$. Hence, there exists $n > 0$ such that $b^{-n}ab^n \wedge a > 1$; let n be the smallest such positive integer.

Suppose that $b^{-n}ab^n$ and a have the same value M. If $Ma \neq Mb^{-n}ab^n$, let H be the ℓ-subgroup generated by M, a and b^n. Since M is the value of a, $b^{-n}Mb^n$ is a value for $b^{-n}ab^n$ and so $b^{-n}Mb^n = M$. M is a normal value, hence $a^{-1}Ma = M$. It follows from Lemma 10.20 that M is an ℓ- ideal of H and H/M is a nonabelian o-group.

If $Ma = Mb^{-n}ab^n$, let $K = \{g \in G : |g| \ll a\}$. Then K is a convex ℓ-subgroup of G, by Lemma 10.21, and $K = M \cap G(a \vee b^{-n}ab^n)$, by the proof of Lemmas 10.21. But then $K = \{g \in G : |g| \ll b^{-n}ab^n\}$ which implies $b^{-n}Kb^n = K$. Let $L = K \boxplus b^{-1}Kb \boxplus b^{-2}Kb^2 \boxplus \ldots \boxplus b^{1-n}Kb^{n-1}$ and let H be the ℓ-subgroup of G generated by L, a and b. Obviously a normalizes K and since a is disjoint from each $b^{-m}Kb^m$ for $m = 1, \ldots, n-1$, it follows that a normalizes L. Clearly b normalizes L and so by Lemma 10.20 L is an ℓ-ideal of H and H/L is ℓ-isomorphic to the Scrimger group G_n, which is a contradiction.

Finally, suppose that $b^{-n}ab^n$ and a have different values. Since $b^{-n}ab^n$ and a are not disjoint but special, their values must be comparable. Without loss of generality we may assume that $b{-}nab^n \gg a$. Let $c = b^n$, M be the value of a and S be the subgroup of G generated by a and c. We now show that for all $t \in S$ that $t^{-1}Mt$ is comparable to M.

For suppose that for some $t \in S$ that $t^{-1}Mt$ is incomparable with M. Then $t^{-1}at \wedge a = 1$. Now $t = a^{m_1}c^{n_1}a^{m_2}c^{n_2}\ldots a^{m_k}c^{n_k}$, where $k \geq 1$ and only m_1 or n_k can be zero. Then

$$\begin{aligned} t^{-1}at &= (c^{-n_k}a^{-m_k}\ldots c^{-n_1}a^{-m_1})a(a^{m_1}c^{n_1}\ldots a^{m_k}c^{n_k}) \\ &= (c^{-n_k}a^{-m_k}c^{n_k})(c^{-n_k-n_{k-1}}a^{-m_{k-1}}\ldots a^{m_{k-1}}c^{n_k+n_{k-1}})(c^{-n_k}a^{nm_k}c^{n_k}), \end{aligned}$$

and continuing in this fashion, we can write $t^{-1}at$ in the form

$$(c^{-s_k}a^{-m_k}c^{s_k})(c^{-s_{k-1}}a^{-m_{k-1}}c^{s_{k-1}})\ldots(c^{-s_1}ac^{s_1})\ldots(c^{-s_{k-1}}a^{m_{k-1}}c^{s_{k-1}})(c^{-s_k}a^{m_k}c^{s_k}).$$

Let $r = \max\{s_1, s_2, \ldots, s_k\}$. Then by Lemma 10.21,

$$g = (c^{-s_2}a^{m_2}c^{s_2})(c^{-s_3}a^{m_3}c^{s_3})\ldots(c^{-s_k}a^{m_k}c^{s_k}) \ll c^{-(r+1)}ac^{r+1}.$$

Now if $s_1 \geq 0$ then

$$a \leq c^{-s_1}ac^{s_1} \ll c^{-(r+1)}ac^{r+1} \text{ and}$$

$$1 \leq g^{-1}ag \wedge a \leq g^{-1}(c^{-s_1}ac^{s_1})g \wedge a = t^{-1}at \wedge a = 1,$$

giving us **Z** wr **Z** by Lemma 10.22.

If $s_1 < 0$, then $c^{-s_1}ac^{s_1} < a$ and so we have that

$$1 \leq g^{-1}(c^{-s_1}ac^{s_1})g \wedge c^{-s_1}ac^{s_1} \leq g^{-1}(c^{-s_1}ac^{s_1})g \wedge a = 1,$$

again giving **Z** wr **Z**. Thus $t^{-1}Mt$ is comparable to M for all $t \in S$.

Therefore the set $\{t^{-1}Mt : t \in S\}$ is a chain of special values. Let $P = \bigcap_{t\in S} t^{-1}Mt$ and H be the ℓ-subgroup of G generated by P and S. Then P is an ℓ-ideal of H and H/P is a nonabelian o-group. //

Gurchenkov [84a] has previously shown that the only covers of $\mathcal{A}$ that contain nonabelian solvable ℓ-groups are the Scrimger varieties and the Medvedev varieties. This result is now a corollary to Theorem 10.23, since Medvedev [77] has shown that the only representable covers of $\mathcal{A}$ that contain a nonabelian solvable ℓ-group are the Medvedev varieties.

Chapter 11: *Groups of Divisibility*

An important application of abelian ℓ-groups is to the theory of integral domains, via groups of divisibility. In this chapter we shall explore the connection between these two theories, which is facilitated by the fact that every abelian ℓ-group can be obtained as a group of divisibility of a suitable domain (see Theorems 11.1 and 11.2 below). These ideas grew out of the classical theory of valuations on fields (see [Ja] for example). Krull [32] was the pioneer in this work, with Jaffard [53] and Ohm [69] making important contributions. Numerous authors have since made use of ℓ-groups to solve ring-theoretic problems; examples of such an approach can be found in Ohm [66], [69], Heinzer [69], Sheldon [73], Hill [72], Lantz [75], Brewer, Conrad and Montgomery [74], and Anderson and Watkins [86]. Valuable books in the field are [M] and [Gi].

Let D be an integral domain with group of units U and quotient field K. If K^* denotes the multiplicative group of K, then U is a subgroup of K^*, and consequently we may define the *group of divisibility* of D as $G(D) = K^*/U$. (We shall suppress the D in this notation unless it is needed for clarity.) Since G is an abelian group, we shall abuse notation and write its operation additively; namely, $Ua + Ub = Uab$.

We can now equip G with a partial order by calling $Ua \leq Ub$ if and only if $b/a \in D$. The reader may easily check that this relation satisfies the reflexive, symmetric and transitive properties, and so is a partial order, and in fact, that this partial order is preserved by the group operation; thus, G is a partially ordered group. Note that the positive cone of G is precisely $\{Ua : a \in D\}$.

Alternatively one could consider the set $\{Dx : x \in K^*\}$ of all nontrivial cyclic D-submodules of K (these are also called the *principal fractional ideals* of D). It is easy to see that this set is a group under the operation $Dx \cdot Dy = Dxy$, which we can make a partially ordered group by setting $Dx \leq Dy$ if and only if $Dx \supseteq Dy$. It is evident that this partially ordered group is o-isomorphic to G under the map $Ux \mapsto Dx$.

We shall now focus our attention on those domains for which the partial order on the groups of divisibility described above is a lattice order. Toward that end, we review a bit of elementary ring theory. Let a and b be elements of a domain D; then c is a *greatest common divisor* (gcd) of a and b if c divides both a and b, and if d is any other common divisor of a and b, then d divides c. A domain is a *pseudo-Bezout* domain, if each pair of elements has a gcd (such domains are also called *GCD-domains*). A domain is a *Bezout domain* if every finitely generated ideal is principal; it is easy to show that a Bezout domain is pseudo-Bezout. In fact, Bezout domains can be characterized as those pseudo-Bezout domains with the additional property that the gcd of any finite set can be expressed as a linear combination of the elements of the set (see [Gi]).

Theorem 11.1 A domain is a pseudo-Bezout domain if and only if the above partial order on its group of divisibility is a lattice order.

Proof: Suppose that D is a pseudo-Bezout domain. To show that $G(D)$ is an ℓ-

group, we need only check that two positive elements Ua and Ub have a meet. We may as well assume that Ua and Ub are strictly positive, that is, that a and b are nonunits. Thus, a and b have a gcd c; it is then evident that Uc is the greatest lower bound for Ua and Ub. The converse follows equally easily. //

In our discussion above we have started with a domain, and obtained its group of divisibility from the quotient field. Alternatively, we can begin with the field, equipped with a function onto an ℓ-group, as follows. Given a field K a *demivaluation* on K is a group homomorphism w from K^* onto an abelian ℓ-group G which satisfies the following property:

$$(*) \qquad w(x+y) \geq w(x) \wedge w(y).$$

If we then let $D_w = \{x \in K : w(x) \geq 0\} \bigcup \{0\}$, it is quite easy to verify that D_w is a pseudo-Bezout domain, whose group of divisibility is ℓ-isomorphic to G in the obvious way.

We will now establish the crucial fact enabling translation of ring-theoretic problems into the language of abelian ℓ-groups, namely, that every abelian ℓ-group arises as the group of divisibility of some domain. Krull [32] proved this for o-groups, and Jaffard [53] generalized his proof to ℓ-groups. (Kaplansky also did this in his thesis but did not publish the result). Ohm [66] observed that the domain obtained by this construction is in fact a Bezout domain.

Theorem 11.2 Every abelian ℓ-group is the group of divisibility of a Bezout domain.

Proof: Let G be an abelian ℓ-group. Given a field K, we shall now consider the *group ring* $K[G]$ of G over K. Now $K[G]$ is defined as the K-vector space with basis consisting of the formal symbols $\{X^g : g \in G\}$, equipped with a multiplication as follows: we first define it for monomials by setting

$$(aX^g) \cdot (bX^h) = abX^{g+h}, \text{ for all } a, b \in K \text{ and } g, h \in G,$$

and then extending this definition using the distributive law to all elements of $K[G]$. It is a standard algebraic exercise to verify that this multiplication is well-defined, and makes $K[G]$ a ring. We will check that $K[G]$ is in fact a domain. For $r \in K[G]$, let $E(r)$ be the set of elements of G which appear as exponents in the (unique) expression for r as a linear combination of X^g's. Given $0 \neq r, s \in K[G]$, we must show that $rs \neq 0$. Equip the torsion-free abelian group G with a compatible total order $\prec$, and suppose that g (respectively, h) is the smallest element of $E(r)$ (respectively, $E(s)$) with respect to $\prec$. But then there will be a unique X^{g+h} term when the product rs is multiplied out, with a necessarily nonzero coefficient, since K is a domain.

Let k be the quotient field of $K[G]$. We shall obtain a demivaluation on k by defining a map w from $K[G]$ onto G^+, which can then obviously be extended to all of k. Given an arbitrary element

$$r = \sum_{i=1}^{n} a_i X^{g_i}$$

of $K[G]$, let $w(r) = g_1 \wedge g_2 \wedge \ldots \wedge g_n$. This map is certainly onto G^+ and so its extension to k will be onto G. We need next to check that w satisfies condition (*) in the definition of demivaluation. For $r, s \in K[G]$, it is evident that $E(r+s) \subseteq E(r) \cup E(s)$, which means that

$$w(r+s) = \bigwedge E(r+s) \geq \bigwedge E(r) \wedge \bigwedge E(s) = w(r) \wedge w(s).$$

Thus condition (*) holds. We must also check that w is a group homomorphism. But because G is abelian and hence representable, we may as well assume that G is an o-group. In this case, $w(r)$ is the minimum element of $E(r)$, and the argument above which shows that $K[G]$ is a domain also shows that $w(rs) = w(r) + w(s)$, for all $r, s \in K[G]$.

We now know that the subring D_w of k is a pseudo-Bezout domain whose group of divisibility is G. It remains only to check that D_w is a Bezout domain. Suppose that $r, s \in D_w$; we will show that the ideal $\langle r, s\rangle$ is principal. Let $w(r) = g$; we then have that $r = tX^g$ where $w(t) = 0$. In a similar way, we have that $s = uX^h$, where $w(s) = h$ and $w(u) = 0$. But then, $\langle r, s\rangle = \langle X^g, X^h\rangle = \langle X^{g\wedge h}\rangle$, which is a principal ideal as claimed. //

Note that this theorem implies that the distinction between pseudo-Bezout domains and Bezout domains is lost upon passage to the group of divisibility. Consequently it will do us no harm to restrict our attention to Bezout domains, especially since this will simplify some of the results and proofs which follow. The reader should note that various of these results certainly do hold in more general contexts.

We shall now establish the correspondence which exists between overrings of a Bezout domain and the convex ℓ-subgroups of its group of divisibility, which will then enable us to easily translate from the language of ℓ-groups to the language of rings, and vice versa.

For this purpose, we need to recall a bit more terminology from the theory of domains. For a domain D, a subset S is a *multiplicative system* if it does not contain 0, and is closed under multiplication; we say that S is *saturated* if it contains all the divisors of elements of S; it is important to note that this condition is usually harmless, since clearly any multiplicative system is contained in a (unique) saturated one. Note that a saturated multiplicative system necessarily contains all the units of the ring. The *localization* of D at S, written D_S, is the subring of the quotient field K of D given by

$$\{d/s \in K : d \in D \text{ and } s \in S\}.$$

Of course, an ideal P is prime exactly if $D\backslash P$ is a multiplicative system; we will follow the usual practice of writing the localization in this case as D_P.

Now a ring R between a domain D and its quotient field K is called an *overring* of D. If D is a Bezout domain, then all overrings of D are actually localizations of D at some (saturated) multiplicative system (see [Gi]); this need not be the case for pseudo-Bezout domains.

Theorem 11.3 Let D be a Bezout domain with quotient field K and $G(D) = K^*/U$ its group of divisibility. Then there are one-to-one correspondences between the following sets:

1) the overrings of D,
2) the saturated multiplicative systems of D,
3) the convex ℓ-subgroups of $G(D)$, and
4) the ℓ-homomorphic images of $G(D)$.

The correspondences between 1) and 2) and between 3) and 4) are the obvious ones. Given a convex ℓ-subgroup H of $G(D)$, the corresponding saturated multiplicative system is $\{d \in D : Ud \in H\}$. Furthermore, given a saturated multiplicative system S of D, the group of divisibility $G(D_S)$ is exactly the ℓ-homomorphic image of $G(D)$ under this correspondence. Finally, under this correspondence the prime subgroups of $G(D)$ correspond exactly to the complements of prime ideals of D.

Proof: Clearly, to establish the one-to-one correspondences, we need only address ourselves to the correspondence between 2) and 3). Given a convex ℓ-subgroup H of $G(D)$, let

$$S = \{d \in D : Ud \in H\};$$

this is a multiplicative system of D. If $s \in S$ and t divides s, then $Ut \leq Us$ and so $t \in S$; thus S is saturated. Conversely if S is a saturated multiplicative system, and we let $H^+ = \{Us \in G : s \in S\}$, then H^+ is a convex subsemigroup of G^+, and as such the positive cone of a convex ℓ-subgroup H of G. It is evident that these maps are inverses of one another.

Now suppose that S is a saturated multiplicative system. Then the group of units of D_S clearly contains U, and is in fact the subgroup V of K^* generated by the subsemigroup S. But then

$$G(D_S) = K^*/V \cong (K^*/U)/(V/U) = G(D)/H,$$

where H is exactly the convex ℓ-subgroup of $G(D)$ to which S corresponds, as described above.

Finally, suppose that $S = D\backslash P$, where P is a prime ideal. Then S corresponds to the convex ℓ-subgroup H of $G(D)$ with positive cone $\{Ud : d \in P\}$. Now suppose that $Ux \wedge Uy = U1$; then 1 is a gcd for x and y and so $1 = ax + by \in \langle x, y\rangle$. Thus both x and y cannot belong to the proper ideal P, and so Ux or Uy belongs to H^+, making H a prime subgroup. Conversely, if H is a prime subgroup of G, we claim that $S = \{d \in D : Ud \in H\}$ is the complement of a prime ideal; clearly, we need only show that the complement is an ideal, since we know that S is a multiplicative system. So if $x, y \in D\backslash S$, then $Ux, Uy \notin H$. Let r be a gcd of x and y; then $Ur = Ux \wedge Uy \notin H$, and since $Ur \leq U(ax + by)$, for all $a, b \in D$, this means that $ax + by \in D\backslash S$. In other words $D\backslash S$ is an ideal. //

The order-reversing correspondence between the prime ideals of a Bezout domain and the prime subgroups of its group of divisibility can be exploited in a number of ways. We will mention a couple of these applications informally.

Of course, the most important historically of these applications is the fact that the ideals of a domain are totally ordered if and only if its group of divisibility is an o-group.

Such domains are *valuation* domains, since they arise from valuations on fields. The original definition of valuation domain asserts that for every element of the quotient field, either it or its inverse belongs to the domain; it is evident directly from the definition of the partial order that the group of divisibility of such a domain is totally ordered. However, we are emphasizing here that the ring property (and the group property) can be looked at in terms of ideals (and convex ℓ-subgroups). We record the essence of this discussion in the following corollary; note that we omit mention of the Bezout hypothesis, since a valuation domain is clearly Bezout.

Corollary 11.4 A domain is a valuation domain if and only if its group of divisibility is a totally ordered group.

Since the group of divisibility of a Bezout domain can be represented as a subdirect product of totally ordered groups, the following classical result [Gi] from domain theory follows immediately:

Corollary 11.5 A Bezout domain is an intersection of overrings which are valuation domains.

If the group of divisibility of a valuation domain has finite rank as an o-group, that is, it is a lexicographic sum of n archimedean o-groups, we say that the domain has *rank* n; in particular, a valuation domain has rank 1 if its group of divisibility is a subgroup of the reals. We shall enquire later into the more general question of when a Bezout domain has an archimedean group of divisibility.

For another application, observe that it is now evident that any *tree* (a partially ordered set where the subset of elements below a given element is totally ordered) occurs as the partially ordered set of primes of a Bezout domain. For the order dual of a tree is a root system, and if Γ is any root system, the ℓ-group $\Sigma(\Gamma, \mathbf{R})$ has (an isomorphic copy of) Γ as its root system of primes. Now apply Theorem 11.1.

Any property of abelian ℓ-groups describable in terms of its primes can now be translated into a property of Bezout domains, and conversely. Thus, a Bezout domain has Krull dimension 0 (that is, every prime ideal is minimal) if and only if its group of divisibility has the property that every prime subgroup is maximal. The latter class of ℓ-groups has been much considered (see Conrad [74c]); it consists precisely of those ℓ-groups for which every ℓ-homomorphic image is archimedean. These are the *hyperarchimedean* ℓ-groups(see class $\mathcal{HA}$ in the appendix).

For another example consider the class of Bezout domains for which each prime is contained in a unique maximal prime; this translates into the class of ℓ-groups with the 'stranded prime' property: each prime contains a unique minimal prime. These ℓ-groups are called *semiprojectable*; see class $\mathcal{B}$ in the appendix. It is well-known that a projectable ℓ-group has the stranded prime property (see [BKW] for a proof); however, the converse is false, as E38 demonstrates. Now the projectability property easily translates in the ring context into the notion of an *adequate* domain (see Larsen, Lewis and Shores [74]); a domain

D is adequate if it is Bezout and for all $0 \neq a, b \in D$, there exist $c, d \in D$ with $a = cd$, $D = (b, c)$ and if s is a nonunit divisor of d, the $(b, s) \neq D$. Conrad, Brewer and Montgomery [74] then used Theorem 11.1 to provide an example of a domain for which each prime is contained in a unique maximal prime, but which is not adequate (see E38).

We shall now turn to the question of describing those Bezout domains whose groups of divisibility are archimedean. This property will of course not be describable in terms of the tree of primes, since being archimedean is not describable in terms of the root system of primes (see E53).

To accomplish this, we need the notion of a completely integrally closed domain. Let D be a domain with quotient field K, and suppose that $0 \neq a, x \in K$. Suppose further that $ax^n \in D$ for all positive integers n implies that $x \in D$; then D is *completely integrally closed*. The following theorem appears essentially in [Gi]:

Theorem 11.6 A Bezout domain is completely integrally closed if and only if its group of divisibility is an archimedean ℓ-group.

Proof: First suppose that D is completely integrally closed. Let Ua, Ub be positive elements of $G(D)$, and suppose that $n(Ua) \leq Ub$, for all positive integers n. Then $Ub + n(U(1/a)) \geq 0$, which means that $b(1/a)^n \in D$, for all positive n. But then $1/a \in D$, and so $Ua = 0$. Hence, $G(D)$ is archimedean.

But then $Ua \geq nU\frac{1}{x}$ for all positive n, and so $U\frac{1}{x} \leq 0$, since $G(D)$ is archimedean. Thus $Ux \geq 0$ and so $x \in D$, as required. //

We can now describe an interesting ring-theoretic interpretation of the Dedekind completion of an archimedean ℓ-group (see Theorem 8.2.2). For that purpose, we need a bit more terminology from ring theory. A *fractional ideal* of domain D is a D-submodule of the quotient field. (Recall in particular that we called the modules Dx principal fractional ideals in the discussion prior to Theorem 11.1.) A fractional ideal I of a domain D is *divisorial* (alternative terminology: a *v-ideal*) if I is the intersection of all principal fractional ideals containing I. Let $Di(D)$ denote the set of all such ideals; it is then evident from our discussion of the set of principal fractional ideals at the beginning of this chapter that $Di(D)$ contains (an isomorphic copy of) $G(D)$. Although we will not pursue the details here, one can then define a multiplication on $Di(D)$ consistent with that of $G(D)$ which makes $Di(D)$ a commutative semigroup; furthermore, $Di(D)$ is a group if and only if D is completely integrally closed (see [Gi]). In fact, $Di(D)$ is then a Dedekind complete ℓ-group, in which $G(D)$ is large (since the order on $Di(D)$ is the order dual of inclusion). Thus, $G(D)$ itself is Dedekind complete if and only if $G(D) = Di(D)$; that is, the group of divisibility of a Bezout domain is Dedekind complete if and only if each divisorial fractional ideal is principal. Domains with the latter property have been called *pseudo-principal* (see [Bo]). Mott [73] has pointed out that this reasoning can be used for a ring-theoretic proof of the fact that every archimedean ℓ-group can be ℓ-embedded into a Dedekind complete ℓ-group.

We can further exploit the correspondence between archimedean ℓ-groups and completely integrally closed domains as follows. First, observe that any ℓ-group admits a unique maximum archimedean ℓ-homomorphic image: For an ℓ-group G, consider the set of normal convex ℓ-subgroups whose corresponding ℓ-homomorphic images are archimedean (this set includes G and so is not empty); since the class of archimedean ℓ-groups is obviously closed under subdirect products, this means that the intersection of this set of convex ℓ-subgroups provides the kernel of the required maximum archimedean ℓ-homomorphic image. (Using the language of Martinez [80b], we're just observing that the class of archimedean ℓ-groups is a torsion-free class) Now, if the ℓ-group under consideration is the group of divisibility $G(D)$ of a Bezout domain D, its maximum archimedean ℓ-homomorphic image provides, by Theorem 11.3, a minimum overring R of D which is completely integrally closed.

Unfortunately, the domain R which we have just constructed need not be the complete integral closure of D, since the complete integral closure need not be completely integrally closed. In fact, the first construction (Heinzer [69]) of an example of a domain where this is not the case involved finding an abelian ℓ-group G for which G/B is not archimedean, where B is the convex ℓ-subgroup of infinitely small elements (see Heinzer [69] or [Gi] for definitions, proofs and a discussion).

A famous conjecture of Krull's [32] was that the complete integral closure of a domain should be the intersection of of the rank 1 valuation overrings in its quotient field; Nakayama [42a], [42b] provided the first counterexample. Note that this conjecture translates into saying that an archimedean ℓ-group is a subdirect product of reals, which we know by E21 is false; see [Gi] for further discussion.

Note that the program carried out above for archimedean ℓ-groups can be generalized to any class of abelian ℓ-groups which is closed with respect to subdirect products, and in particular, any torsion-free class; that is, such a class gives rise to a corresponding class of domains, such that every domain admits a unique minimal overring belonging to the class (see Anderson [86]).

Appendix: *A Menagerie of Examples*

In this appendix we shall present a collection of examples of lattice-ordered groups to assist the reader in exploring the boundaries of the theory sketched in the previous chapters. We shall for the most part omit proofs, referring the interested reader to other parts of the text or to the literature.

We shall begin by listing a number of the most important classes of ℓ-groups, with particular emphasis on varieties and torsion classes. In the process of describing these classes we shall introduce further terminology from the subject, which we will use in classifying the examples per se.

Section 1: *Varieties of ℓ-groups*

The following are descriptions of some important varieties. Many of these have been more fully described in the text, particularly in Chapter 7. A diagram of the inclusion relations among them follows. The literature on varieties of ℓ-groups is large and growing; an interesting survey by an important worker in the field is Holland [83].

$\mathcal{E}$: The trivial variety

This variety consists only of the trivial one element ℓ-group, and is defined by the law $x = 1$.

$\mathcal{A}$: The abelian variety

The class of all abelian ℓ-groups is the smallest proper variety of ℓ-groups (see Section 7.1); since it is defined by a single law, it is not the intersection of a descending chain of varieties, and hence has the covers described in Section 7.2. Also, this variety is the smallest variety containing the ℓ-group $\mathbf{Z}$.

$\mathcal{A}^n$: The abelian powers

Note that the members of this ascending chain of the powers of the abelian variety are all distinct (Lemma 7.18), and that their join in the lattice $\mathbf{V}$ of varieties is the normal-valued variety $\mathcal{N}$. Note however that the smallest complete torsion class $\mathcal{A}^*$ which contains the abelian powers is smaller than $\mathcal{N}$ (see E44).

$\mathcal{R}$: The representable variety

This variety is considered in Chapter 4, where we show that it is defined by a single equation (Theorem 4.1.1). Note that of course this variety is generated by the class of all o-groups.

$\mathcal{N}$: The normal-valued variety

The variety of all normal-valued ℓ-groups is the largest proper variety (Theorem 7.1.4). It is the only idempotent variety (aside from the trivial varieties $\mathcal{L}$ and $\mathcal{E}$; see Corollary 7.1.12). Consequently, it is the join of the ascending powers of any nontrivial variety. In contrast, it is a complete torsion class which as a torsion class is strongly "join inaccessible"; see the discussion in Holland and Martinez [79].

$\mathcal{L}$: The variety of all ℓ-groups

Note that this class is not a variety of groups, since subgroups of ℓ-groups need not admit a lattice order (see E2); indeed, there exists no satisfactory group-theoretic characterization of the groups belonging to this class (see the discussion of right-ordered groups at the end of Chapter 5).

$\mathcal{L}_n$: The variety of commuting powers

These varieties are generated by the laws $x^n y^n = y^n x^n$. Martinez [72] showed that $\mathcal{L}_n \cap \mathcal{R} = \mathcal{A}$. Consider the following example of an ℓ-group in $\mathcal{L}_p \backslash \mathcal{S}_p$ for prime p due to McCleary, found in Fox [83]: Let T be the subgroup of $\mathbf{R}$ generated by 1 and $\sqrt{2}$ and $pT = \{pt : t \in T\}$. Denote by H_p the ℓ-subgroup of $\mathbf{Z}$ Wr T consisting of the elements $(\langle h_i \rangle, \bar{h})$ where $i \equiv j \pmod{pT}$ implies that $x_i = x_j$.

Gurchenkov [84b] has shown that each of the $\mathcal{L}_p$, for p prime, is contained $\mathcal{A}^2$. Reilly [86] has shown that if n is the product if k primes then $\mathcal{L}_n \subseteq \mathcal{A}^{k+1}$ and in addition that if $\mathcal{V} \cap \mathcal{R} = \mathcal{A}$ then $\mathcal{V} \subseteq \mathcal{L}_n$ for some n. Holland, Mekler and Reilly [86] contains other recent results concerning these varieties.

$\mathcal{S}_n$: The Scrimger varieties

These varieties have been considered in some detail in Section 7.2. Recall that for each positive integer $n > 2$, $\mathcal{S}_n$ is generated by the ℓ-group G_n where $G_n \subseteq \mathbf{Z}$ Wr $\mathbf{Z}$ is defined by:

$$G_n = \{(\langle w_i \rangle, \bar{w}) : i \equiv j \pmod{n} \text{ implies } w_i = w_j\}.$$

Each $\mathcal{S}_n \subseteq \mathcal{L}_n$ and by Theorem 10.23 the Scrimger varieties are the only nonrepresentable covers of $\mathcal{A}$. We show in Section 7.2 that $\mathcal{S}_n \subseteq \mathcal{L}_n$. The group H_p defined above is in $\mathcal{L}_p$ but not in $\mathcal{S}_p$ for prime p (see Fox [83]). Hence $\mathcal{S}_p \subset \mathcal{L}_p$. A similar containment exists for composite n, as the Smith varieties demonstrate.

$\mathcal{S}_{n_1,\ldots,n_k}$: The Smith varieties

Smith [81] has given a generalization of the Scrimger varieties . Notice that each ℓ-group G_n defining the Scrimger varieties can also be considered as a splitting extension of the form

$$(\prod_0^{n-1} \mathbf{Z}) \overleftarrow{\times}_\theta \mathbf{Z}$$

where $\theta : \mathbf{Z} \to Aut(\prod_0^{n-1} \mathbf{Z})$ affects the group operation by

$$((a_0, \ldots, a_{n-1}), b)((c_0, \ldots, c_{n i 1}), d) = ((a_0 + c_b, \ldots, a_{n-1} + c_{n-1+b}), b + d)$$

where all subscripts are modulo n.

Now define inductively an ℓ-group

$$G_{n_1,\ldots,n_k} = (\prod_0^{n_k - 1} G_{n_1,\ldots,n_{k-1}}) \overleftarrow{\times}_\theta \mathbf{Z}$$

for any finite sequence $n_1, \ldots, n_k \in \mathbf{N}\backslash\{1\}$. (Such a ℓ-group could also be viewed as an ℓ-subgroup of $Wr^{k+1}\mathbf{Z}$ with the appropriate restrictions at each level, as the Scrimger groups were originally given.) Let $\mathcal{S}_{n_1,\ldots,n_k}$ be the ℓ-group variety generated by $G_{n_1,\ldots,n_k}$.

Smith has shown that

a) $G_{n_1,\ldots,n_k} \in \mathcal{L}_n$ iff $(n_1 \ldots n_k)|n$.

b) For any $k \in \mathbf{N}$ and any finite sequence $n_1, \ldots, n_k \in \mathbf{N}\backslash\{1\}$, $G_{n_1,\ldots,n_k} \in \mathcal{A}^{k+1}\backslash\mathcal{A}^k$.

c) For any $k \in \mathbf{N}$ and any infinite set $\{n_1(\lambda), \ldots, n_k(\lambda) : \lambda \in \Lambda\}$ in $\mathbf{N}^k$ which contains an infinite subset $\{n_1(\mu), \ldots, n_k(\mu) : \mu \in M\}$ for which $n_i(\mu) \neq n_i(\nu)$ for all $1 \leq i \leq k$ whenever $\mu \neq \nu$,

$$\bigvee_{\lambda \in \Lambda} \mathcal{S}_{n_1(\lambda),\ldots,n_k(\lambda)} = \mathcal{A}^{k+1}.$$

(It follows that $\bigvee_{n=2}^{\infty} \mathcal{S}_n = \mathcal{A}^2$.)

$\mathcal{H}$: A "large" variety properly contained in $\mathcal{R}$

Define the variety $\mathcal{H}$ to be those representable ℓ-groups satisfying the equation

$$(*) \qquad \big|[b, |[a,b]|]\big| \ll b \text{ where } e \leq b \leq a.$$

Let $\bar{\mathcal{A}}^n = \mathcal{A}^n \cap \mathcal{R}$, the "representable abelian powers". Then $\bar{\mathcal{A}}^n \subseteq \mathcal{H}$ for all $n \in \mathbf{N}$. However, $\mathcal{H} \subset \mathcal{R}$ since Clifford's o-group (E9) does not satisfy (*). Hence $\bigvee \bar{\mathcal{A}}^n \subset \mathcal{R}$.

This shows that the lattice of ℓ-group varieties is not Brouwerian: Since $\bigvee \mathcal{A}^n = \mathcal{N}$ and $\mathcal{R} \subseteq \mathcal{N}$, $\mathcal{R} \wedge (\bigvee \mathcal{A}^n) = \mathcal{R}$. However, $\mathcal{R} \cap \mathcal{A}^n = \bar{\mathcal{A}}^n$ and $\bigvee \bar{\mathcal{A}}^n \subset \mathcal{R}$. Smith [81] first showed that the lattice of ℓ-group varieties is not Brouwerian using the Scrimger varieties.

$\mathcal{M}^+, \mathcal{M}^-, \mathcal{N}_0$: The Medvedev varieties

Recall from Section 7.2 that $\mathcal{M}^+$ is obtained by totally ordering $\mathbf{Z}$ wr $\mathbf{Z}$ by calling an element $g = (\langle g_i \rangle, \bar{g})$ positive if $\bar{g} > 0$ or $\bar{g} = 0$ and $g_i > 0$ where i is maximal in the support of of $\langle g_i \rangle$. $\mathcal{M}^-$ is defined similarly except an element is positive if $\bar{g} > 0$ or $\bar{g} = 0$ and $g_i > 0$ where i is minimal in the support of $\langle g_i \rangle$. $\mathcal{N}_0$ is generated by the free nilpotent class two group

$$N = \langle a, b, c : c = [a,b], [a,c] = 1 = [b,c] \rangle$$

where an element $a^m b^n c^k$ is positive if $m > 0$, $m = 0$ and $n > 0$ or $m = n = 0$ and $k > 0$. All three are covers of $\mathcal{A}$. In fact, any variety which contains a nonabelian solvable o-group must contain one of these three varieties.

$\mathcal{U}_r$ for $0 < r \leq 1$: The Feil varieties

Recall from Section 7.3 that for $\frac{p}{q} \leq 1, p, q \in \mathbf{Z}^+$, $\mathcal{U}_{p/q}$ are those representable varieties defined by $p\big|x, |[x,y]|\big| \geq q\big|[x,y]\big|$. For $r \in \mathbf{R}\backslash\mathbf{Q}, 0 < r < 1$, define $\mathcal{U}_r = \bigcap\{\mathcal{U}_{p/q} : \frac{p}{q} > r\}$. Then $\mathcal{U}_r \subseteq \mathcal{U}_s$ if and only if $r \leq s$. This gives a tower of varieties of order type of the continuum. Note that the intersection of this tower of varieties can not be described by a single equation. Also, the only cover of $\mathcal{A}$ contained in this intersection is $\mathcal{M}^+$. In addition, $\{\mathcal{U}_r \cap \mathcal{A}^2 \cap \mathcal{R}\}$ is also a tower. Denote these varieties by $\bar{\mathcal{U}}_r$.

There is also a "sister" tower of varieties $\mathcal{U}^-_{p/q}$ defined by the equations

$$p\big|[x^{-1},|[x,y]|]\big| \geq q|[x,y]|,$$

along with the corresponding $\mathcal{U}^-_r$ and $\bar{\mathcal{U}}^-_r$.(See Feil [80a].) Then $\mathcal{U}^-_r \subseteq \mathcal{U}^-_s$ if and only if $s \leq r$. The only cover of $\mathcal{A}$ contained in the intersection of the $\mathcal{U}^-_r$ is $\mathcal{M}^-$.

Huss [84a] has observed that the collection of varieties $\{\mathcal{U}_t \vee \mathcal{U}^-_t\}$ is pairwise incomparable.

Rei: **The collection of Reilly varieties**

Reilly [81] has constructed a continuum antichain of varieties all of which contain $\mathcal{R}$. See the end of Section 7.3 for a brief description.

Spec: **The specially ordered variety**

Bergman [84] has shown that a free group can be "specially ordered": $1 < b \ll a$ implies that $b \ll b^a$. Any specially ordered o-group satisfies an equation not satisfied by any of the Medvedev groups; hence there exists another representable cover of $\mathcal{A}$. (See Section 7.2.) Powell and Tsinakis [86] have shown that the two-generator free group equipped with a special order in fact generates a cover of $\mathcal{A}$. We call this cover *Spec*. Using techniques similar to Bergman's one can show that any free group can be "reverse-specially ordered": $1 < b \ll a \Rightarrow b \gg b^a$. Such o-groups satisfy a similar equation to the one mentioned for specially ordered o-groups, which is not satisfied by any of the Medvedev groups nor the specially ordered groups, thus implying the existence of another cover of $\mathcal{A}$. Powell and Tsinakis have also shown that the two-generator free group equipped with a reverse-special order generates another cover of $\mathcal{A}$. We denote this variety by $Spec^-$. Kopytov [85] has independently shown the existence of a reverse-specially ordered o-group.

$\mathcal{W}$: **The weakly abelian variety**

An ℓ-group is *weakly abelian* if for all $a \in G^+$ and $g \in G$, $a + a \geq -g + a + g$. Such ℓ-groups have been considered by Martinez [72b] and Reilly [83]. In particular, Martinez showed that the following are equivalent:

1. G is weakly abelian.

2. $|[x,y]| \ll x$, for all $0 < x \in G$ and $y \in G$.

3. If $0 < a \in G$ and P is a value of a in G with cover P^*, then P is normal in P^* and P^*/P is central in G/P.

a. Evidently, from 2. the collection $\mathcal{W}$ of all weakly abelian ℓ-groups is a variety.

b. Martinez [72b] also proved that all weakly abelian ℓ-groups are Hamiltonian (that is, all convex ℓ-subgroups are normal: see the class *Ham* below). In fact, the variety of weakly abelian ℓ-groups is the largest variety of Hamiltonian ℓ-groups. Since each Hamiltonian ℓ-group is obviously representable, so is each weakly abelian ℓ-group.

c. Note that the varieties $\mathcal{A}^2 \cap \mathcal{R}$ and $\mathcal{W}$ are incomparable: for $\mathbf{Z}$ wr $\mathbf{Z}$ ordered lexicographically is in $\mathcal{A}^2 \cap \mathcal{R}$ while not in $\mathcal{W}$. On the other hand, it is easy to see that the usual total order on any free group F (see E4) makes it weakly abelian and so

$$F \in \mathcal{W} \backslash \mathcal{A}^2 \cap \mathcal{R}.$$

d. For another example of a weakly abelian ℓ-group which is not abelian, see E41. The ℓ-group described there in fact belongs to the smaller variety $\mathcal{A}^2 \cap \mathit{Nil}$ described below.

e. Note that Reilly [83] has proved that each nilpotent ℓ-group is weakly abelian, thus generalizing Hollister and Kopytov's Theorem 4.1.2.

Nil: **The variety of nilpotent ℓ-groups**

Consider the class *Nil* of ℓ-groups which are nilpotent as groups; since the class of nilpotent groups is a group variety, this is obviously a variety of ℓ-groups.

Hollister [78] and Kopytov [82] have proved that nilpotent ℓ-groups are representable; Reilly [83] extended this result by showing that nilpotent ℓ-groups are in fact weakly abelian (see class $\mathcal{W}$ above).

$\mathcal{A}^2 \cap \mathit{Nil}$: **$\ell$-groups with commuting conjugates**

Consider the class of ℓ-groups such that every element commutes with each of its conjugates; this is clearly a variety, which has been considered by Martinez [72b] and Reilly [83].

a. Reilly [83] has shown that this class coincides with those ℓ-groups whose group structure is nilpotent of class at most two. Since every nilpotent ℓ-group is weakly abelian (see the discussion of class $\mathcal{W}$ above), so is every ℓ-group with commuting conjugates. This means that we may denote this class by $\mathcal{A}^2 \cap \mathit{Nil}$.

b. See E41 for an ℓ-group with commuting conjugates which is not abelian.

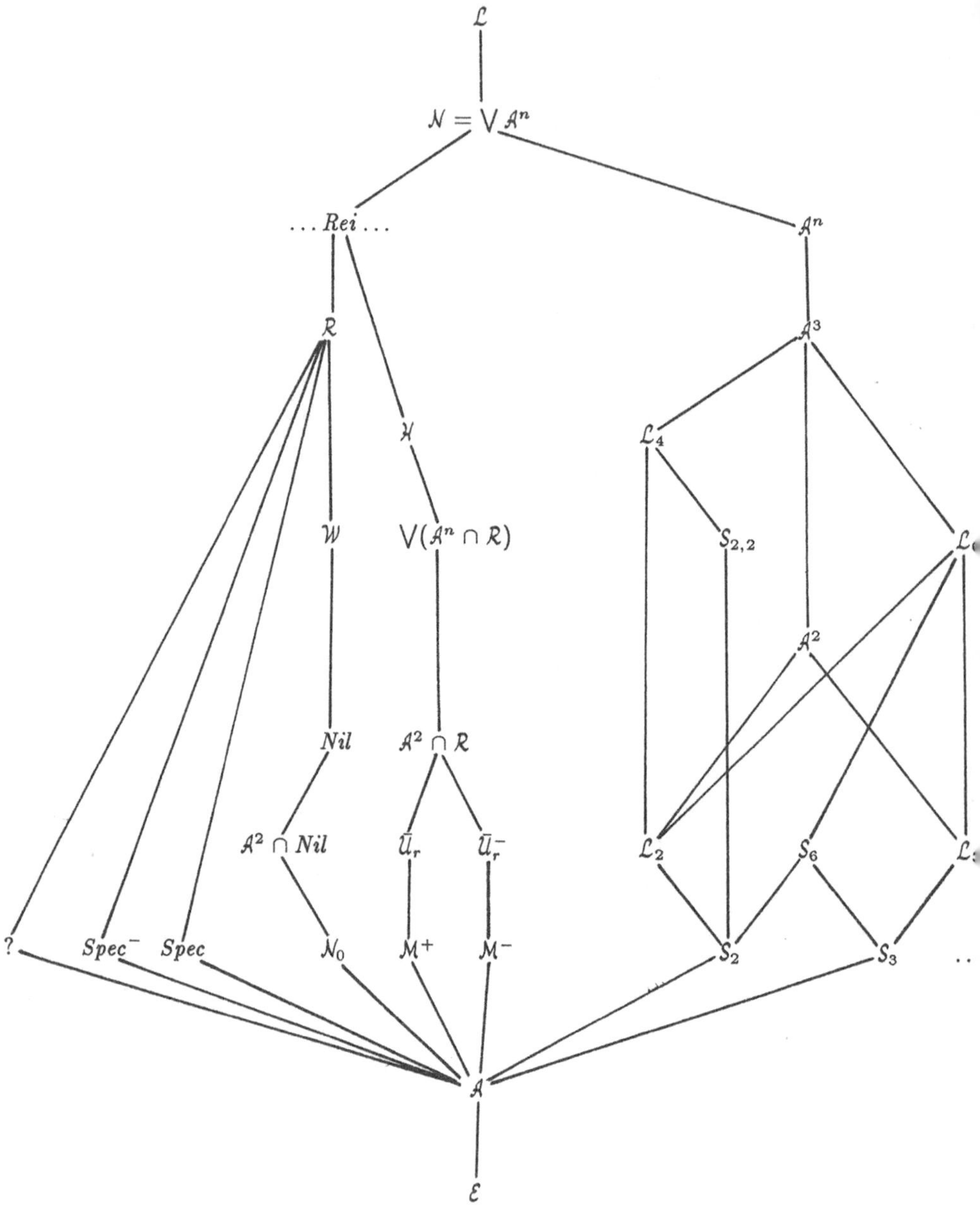
$\mathcal{L}$
$\mathcal{N} = \bigvee \mathcal{A}^n$
... Rei ...
$\mathcal{A}^n$
$\mathcal{R}$
$\mathcal{A}^3$
$\mathcal{H}$
$\mathcal{L}_4$
$\mathcal{W}$
$\bigvee(\mathcal{A}^n \cap \mathcal{R})$
$\mathcal{S}_{2,2}$
$\mathcal{A}^2$
Nil
$\mathcal{A}^2 \cap \mathcal{R}$
$\mathcal{A}^2 \cap Nil$
$\bar{\mathcal{U}}_r$
$\bar{\mathcal{U}}_r^-$
$\mathcal{L}_2$
$\mathcal{S}_6$
?
$Spec^-$
Spec
$\mathcal{N}_0$
$\mathcal{M}^+$
$\mathcal{M}^-$
$\mathcal{S}_2$
$\mathcal{S}_3$
$\mathcal{A}$
$\mathcal{E}$

Section 2: *Torsion and Radical Classes of ℓ-groups*

Many of the desirable structural properties which ℓ-groups might possess are not definable by equations (we shall list numerous examples of such classes); this situation is in contrast to that in group theory (see [Ne], e.g.). There have been several recent attempts to provide a general framework for ℓ-group classification, and we shall define here the most important of these, namely, torsion classes of ℓ-groups. We also define another, radical classes of ℓ-groups. It should be mentioned, however, that unlike the case for varieties, these other classification schemes are for the most part purchased at the cost of allowing a proper class of classes to come under consideration.

A class closed under 1) convex ℓ-subgroups, 2) ℓ-homomorphisms and 3) joins of convex ℓ-subgroups is a *torsion class*. By 3) we mean the following: suppose that $\{C_\alpha\}$ is a collection of convex ℓ-subgroups of an arbitrary ℓ-group G, such that each C_α belongs to the class; then the convex ℓ-subgroup $\bigvee C_\alpha$ of G belongs to the class. This is Martinez's [75b] original definition; he later [80b] dropped property 1) from his definition, saying that the class is *hereditary* if it is also closed under 1). In this appendix we are primarily concerned with hereditary classes, and so we will adopt the original definition. A *quasi-torsion class* is a class closed under 1) and 3) above and also under complete ℓ-homomorphisms.

A class of ℓ-groups closed only under 1) and 3) is a *radical class*; such classes have been considered by Jakubík [77] and Darnel [86a]. A radical class for which membership in the class is determinable within the lattice of closed convex ℓ-subgroups is a *K-radical class*; they have been considered by Conrad [81] and Jakubík [83a].

We shall now describe some important examples of radical and torsion classes which appear in the literature, and give in the diagram below the inclusion relations among them; we shall also include the relations among these classes and the abelian, representable and normal-valued varieties $\mathcal{A}$, $\mathcal{R}$ and $\mathcal{N}$. In our descriptions of these classes we shall refer to the various ℓ-group examples necessary to see that these inclusions are strict. Good general references for a detailed description of these classes are Conrad [77], [81] , Martinez [75b] and Darnel [86a].

$\mathcal{AO}$: Small sums of archimedean o-groups

Let $\mathcal{AO}$ be the class of small sums of archimedean o-groups. This is a torsion class considered in Conrad [77].

$\mathcal{HA}$: Hyperarchimedean ℓ-groups

Let $\mathcal{HA}$ be the class of all ℓ-groups each of whose ℓ-homomorphic images is archimedean. Such ℓ-groups are called *hyperarchimedean*; we mentioned them in Chapter 11. A comprehensive description of their properties may be found in Conrad [74c] (where they are called epi-archimedean); this is a torsion class.

a. Note that an ℓ-group is hyperarchimedean if and only if each prime is minimal, or equivalently, that each principal convex ℓ-subgroup is a cardinal summand.

b. For an example of a hyperarchimedean ℓ-group not belonging to $\mathcal{AO}$ see E15.

Arch: Archimedean ℓ-groups

Let *Arch* denote the class of all archimedean ℓ-groups. We have considered this class extensively in Chapter 2 and Section 8.2. This is a quasi-torsion class (see Kenny [75a]) and a K-radical class (see Conrad [81]).

For an example of an ℓ-homomorphic image of an archimedean ℓ-group which is not archimedean, see E14; more generally, note that the free abelian ℓ-group on a set of generators is archimedean (Theorem 6.4), and so every abelian ℓ-group is the ℓ-homomorphic image of some archimedean ℓ-group.

$\mathcal{O}$: Small sums of o-groups

Let $\mathcal{O}$ be the class of small sums of o-groups. This is a torsion class (Conrad [77]) which obviously contains $\mathcal{A}\mathcal{O}$.

$\mathcal{F}$: Orthofinite ℓ-groups

Let $\mathcal{F}$ be the class of orthofinite ℓ-groups (equivalent terminology: ℓ-groups with *property F*): an ℓ-group belongs to this class exactly if every bounded pairwise disjoint set of positive elements is finite. This is torsion class considered in Conrad [77]. Note that every orthofinite ℓ-group has a basis and is finite-valued (see classes *Fv* and *Bas* below).

a. For a useful class of examples of abelian orthofinite ℓ-groups, see E32.

b. If an ℓ-group is orthofinite, then clearly each convex ℓ-subgroup which is bounded by some element has a finite basis. However, the converse is false; see E19.

Fv: Finite-valued ℓ-groups

Let *Fv* be the class of finite-valued ℓ-groups (which were considered in some detail in Chapter 10); this is a torsion class.

E33 and E42 provide finite-valued ℓ-groups which are not orthofinite.

Bas: ℓ-groups with a basis

Let *Bas* be the class of all ℓ-groups with a *basis*: an ℓ-group has a basis if it has a maximal pairwise disjoint set of positive elements, whose principal convex ℓ-subgroups are o-groups; these latter elements are called *basic*. Such ℓ-groups have been mentioned in passing in the text (see for example the discussion at the end of Section 4.1). This is a K-radical class, but not a torsion class.

a. An element g is basic if and only if the polar g' is prime (see [C]). Thus if an ℓ-group has a basis it has a collection of closed primes whose intersection is zero, and thus is completely distributive. In fact, a refinement of this argument shows that if an ℓ-group has a basis, then its Conrad radical is zero, and so has a minimal plenary subset.

b. E14 is an ℓ-group with a basis with an ℓ-homomorphic image without a basis.

c. E42 is an ℓ-group with a basis which is not representable.

d. E47 provides an ℓ-group with a basis which is not normal-valued; in fact ℓ-groups are constructed there which have even essential values which are not normal in their covers.

$\mathcal{S}$: Special-valued ℓ-groups

Let $\mathcal{S}$ be the class of all special-valued ℓ-groups, which have been considered in Chapter 10. This is a quasi-torsion class, but not a torsion class (see Conrad [81] and Bixler and Darnel [86]).

a. E14 is a special-valued ℓ-group with an ℓ-homomorphic image which is not special-valued.

b. For an ℓ-group which is special-valued but not finite-valued we need only consider a Hahn group on a root system with an infinite pairwise disjoint set.

c. E19 is a normal-valued (and in fact abelian) ℓ-group which is not special-valued.

d. Note that an ℓ-group is normal-valued if and only if it can be ℓ-embedded into a special-valued ℓ-group (see Corollary 10.17) and can be ℓ-embedded in such a way that any normal plenary set of values is mapped to the plenary set of special values. For an example which shows that plenary set must be normal, see E46.

Ess: ℓ-groups with no proper plenary subsets

Let *Ess* be the class of ℓ-groups which have no proper plenary subsets; equivalently, every value is essential. Thus, by Theorem 10.6 every value is closed, and consequently every convex ℓ-subgroup is closed. For normal-valued ℓ-groups, these conditions are then equivalent; presently no example is known of a non-normal-valued ℓ-group with every convex ℓ-subgroup closed, which has non-essential values. E38 is an ℓ-group belonging to *Ess* which is not finite-valued.

Min: ℓ-groups with a minimal plenary subset

Let *Min* be the class of ℓ-groups with a minimal plenary set of values, or equivalently, ℓ-groups whose Conrad radicals are zero (see Chapter 10). This is a K-radical class.

a. Theorem 10.8 asserts that each element of *Min* is completely distributive, and that the converse is true for normal-valued ℓ-groups.

b. E38 is a normal-valued ℓ-group belonging to *Min* which is not special-valued.

c. A full Hahn group $V(\Gamma, \mathbf{R})$ with an infinite root system is an ℓ-group belonging to *Min* but not *Ess*.

$\mathcal{CD}$: Completely distributive ℓ-groups

Let $\mathcal{CD}$ be the class of completely distributive ℓ-groups (which are discussed in Chapter 10). This is a K-radical class (see Conrad [81]).

a. E51 (namely, $A(\mathbf{R})$) is a completely distributive ℓ-group which is not normal-valued, and which does not possess a minimal plenary subset, and in particular does not have a basis.

b. E23 is a completely distributive ℓ-group with an ℓ-homomorphic image which isn't.

$\mathcal{D}$: ℓ-groups with the DCC

Let $\mathcal{D}$ be the class of ℓ-groups which satisfy the descending chain condition on regular subgroups; equivalently, each prime subgroup is regular (Pedersen [70]). This is a complete torsion class (Conrad [77]).

a. The vector lattices belonging to $\mathcal{D}$ have been characterized in Anderson, Bixler and Conrad [83] as those vector lattices which admit no proper a-subspaces (that is, the vector lattice is an a-extension of the subspace).

b. E53 is an element of $\mathcal{D}$ which is not normal-valued.

$\mathcal{A}^*$: The completion of the abelian torsion class

Let $\mathcal{A}^*$ denote the completion of the torsion class $\mathcal{A}$ of all abelian ℓ-groups.

a. E8 is a representable ℓ-group which does not belong to $\mathcal{A}^*$.

b. E42 belongs to $\mathcal{A}^*$ but is not representable.

$\mathcal{SP}$: Strongly projectable ℓ-groups

Let $\mathcal{SP}$ be the class of strongly projectable ℓ-groups, that is, ℓ-groups for which each polar is a cardinal summand; we have discussed such ℓ-groups in sections 1.2 and 8.1.

This class is a quasi-torsion class but not a torsion class (see Anderson [77]); E36 is a strongly projectable ℓ-group with an ℓ-homomorphic image which is not.

$\mathcal{P}$: Projectable ℓ-groups

Let $\mathcal{P}$ be the class of projectable ℓ-groups, that is, ℓ-groups for which each principal polar is a cardinal summand (we discussed such ℓ-groups briefly in Section 8.1).

a. This class is not even a quasi-torsion class; E37 is a projectable ℓ-group with a complete ℓ-homomorphic image which is not projectable.

b. Since all polars (and hence minimal primes) are normal, projectable ℓ-groups are representable.

c. Every prime of a projectable ℓ-group contains a unique minimal prime (Bigard [BKW]); thus each projectable ℓ-group belongs to the class $\mathcal{B}$ (discussed below).

d. If an ℓ-group is projectable then both its lattice of principal polars g'' and its lattice of coprincipal polars g' are sublattices of $\mathcal{C}(G)$ (see Anderson, Conrad and Martinez [86]); however, E38 shows that the converse is false.

e. E19 is an ℓ-group which is projectable but not strongly projectable.

$\mathcal{B}$: Semiprojectable ℓ-groups

Let $\mathcal{B}$ be the class of all ℓ-groups for which each prime contains a unique minimal prime; such ℓ-groups are said to be semiprojectable, or to have the "stranded primes" property. This is a torsion class (Conrad [77]).

a. E38 is an ℓ-group which is semiprojectable but not projectable.

b. E56 shows that the longstanding conjecture that a semiprojectable ℓ-group need be normal-valued is false.

Ham: Hamiltonian ℓ-groups

Let *Ham* be the class of ℓ-groups for which each convex ℓ-subgroup is normal; we call such ℓ-groups *Hamiltonian*. Conrad [80a] proved that *Ham* is a torsion class. This class has also been considered by Martinez [72b] and Reilly [83].

a. Since every minimal prime of a Hamiltonian ℓ-group is normal, it is obvious by Theorem 4.1.1 that each such ℓ-group is representable. Of course, examples abound of representable ℓ-groups which are not Hamiltonian.

b. Martinez [72b] has observed that the variety $\mathcal{W}$ of weakly abelian ℓ-groups is a subclass of the Hamiltonian ℓ-groups; it is in fact the largest variety contained in *Ham* (Reilly [83]). See E43 for a Hamiltonian ℓ-group which is not weakly-abelian.

c. To see that *Ham* is not a variety, see E30.

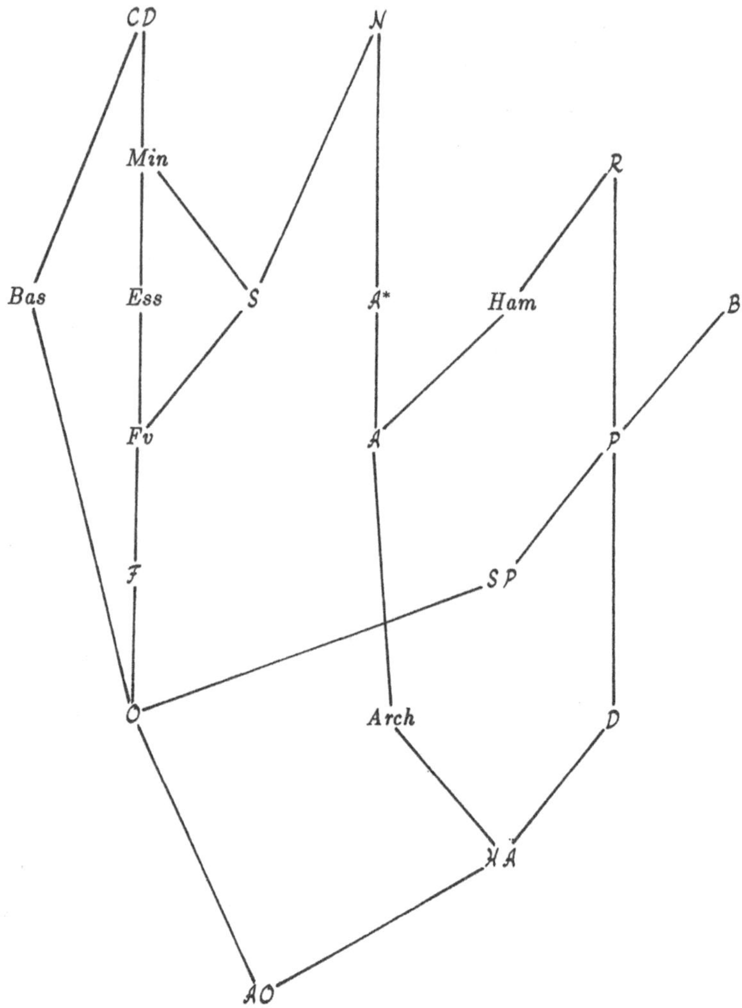

CD
N
Min
R
Bas
Ess
S
A*
Ham
B
Fv
A
P
F
SP
O
Arch
D
HA
AO

Section 3: *Examples of Lattice-ordered Groups*

In this section we shall present a list of examples of ℓ-groups. The examples are given numbers in the form E# for ease of reference. We have made some effort to group similar examples together, starting with archimedean ℓ-groups, followed by abelian, normal-valued and finally non-normal-valued examples. However, being a menagerie of lattice-ordered groups, we can at best expect this section to be partially ordered. The reader should beware that many of the statements about these examples are not trivial to prove; consult the appropriate reference when necessary.

E1: A lattice-orderable group which cannot be totally ordered

Let G and H be any two non-trivial ℓ-groups. Then H Wr G cannot be totally ordered: Pick $h \in H^+$ and $g \in G^+$. Let $w = (\langle w_t \rangle, \bar{w})$ be the element of H Wr G with $\bar{w} = 1$, $w_t = h$ if t is an odd power of g, $w_t = h^{-1}$ if t is an even power of g, and $w_t = 1$ elsewhere. Let $v = (\langle v_t \rangle, \bar{v})$ be the element of H Wr G with $v_t = 1$ for all t and $\bar{v} = g$. Then $v^{-1}wv = w^{-1}$ which impossible for an o-group. Neumann [49] first observed that such large wreath products cannot be totally ordered.

E2: A right-orderable group which cannot be lattice-ordered

Consider the large wreath product **Z** Wr **Z** and the elements v and w of this group as defined in E1 (except, of course, additive notation is used here); recall that w is conjugate to its inverse via v. Let G be the subgroup of the wreath product generated by v and w. Since G is a subgroup of an ℓ-group it can clearly be right-ordered.

Suppose that G could be lattice-ordered. Then

$$w \vee 0 = mv + nw, \text{ for some integers } n, m.$$

Thus

$$mv + (n-1)w = mv + nw - w = (w \vee 0) - w =$$
$$0 \vee -w = 0 \vee (-v + w + v) = -v + (w \vee 0) + v =$$
$$-v + mv + nw + v = mv + -v + nw + v = mv - nw.$$

Consequently $(n-1)w = -nw$, which is a contradiction, since this group is torsion free. This proof is due to Charles Holland [C].

E3: A torsion-free group which cannot be right-ordered

The following example is due to Smirnov [67]; it also appears in Mura and Rhemtulla [MR]. Let

$$G = \langle u, v, b : [u, v, u] = [u, v, v] = 1, u^b = u^{-1}\ ,\ v^b = v^{-1}\ ,\ [u, v] = b^{-16} \rangle.$$

E4: A group whose only lattice orders are total orders

Let F be a free group (on a set of generators of any cardinality). Then the descending series

$$F \supseteq F_1 = [F, F] \supseteq F_2 = [F_1, F_1] \supseteq \cdots$$

terminates after countably many steps in $\{1\}$ and each quotient F_n/F_{n+1} is a torsion-free abelian group (see for example [Ja]). By Proposition 1.1.6 we can choose a total order for each of these quotients. Then totally order F as follows: call an element positive if it is positive in the first of these quotients for which it is nontrivial. It is easily verified that this makes F an o-group. Note that there are total orders on F which do not arise in this fashion (see E56).

We now observe that F cannot admit a lattice order which is not total. For if such existed, there would exist nontrivial elements where $a \wedge b = 1$. But then a and b would commute, while not being in the same cyclic subgroup; this obviously contradicts the freeness of F.

E5: A subdirect product of ℓ-groups which is not lattice-orderable

Let G_2 be the Scrimger 2-group (see Section 7.2); note that this ℓ-group can be conveniently represented as $\mathbf{Z} \times \mathbf{Z} \times \mathbf{Z}$, where the first component corresponds to the top in the wreath product, the second component gives the even terms and the third component the odd terms in the wreath product. This means that $(a,b,c) \geq (0,0,0)$ if either $a > 0$ or $a = 0$ and $b, c \geq 0$; and the group operation is a semidirect product given as follows:

$$(x,y,z)(a,b,c) = \begin{cases} (x+a, y+b, z+c), & \text{if } a \text{ is even} \\ (x+a, z+b, y+c), & \text{if } a \text{ is odd} \end{cases}$$

Now consider H, the unrestricted direct product of countably many copies of G_2, indexed by the positive integers. Let $a = (1,1,-1) \in G_2$ and $b = (1,-1,1)$, and consider $\vec{a}$ and $\vec{b}$ the elements of H which have a and b, respectively, at each coordinate. Then let G be the subgroup of H generated the small sum together with $\vec{a}$ and $\vec{b}$.

a. Thus G is a subdirect product of lattice-orderable groups; another way to express this is to say that G is *residually* lattice-orderable.

b. However Holland [85c] shows that G itself is not lattice-orderable. He does this by observing that $\vec{a}^2 = \vec{b}^2$ but that $\vec{a}$ and $\vec{b}$ are not conjugates; Byrd, Lloyd, Mena and Teller [77] have proved that if the squares of two elements of an ℓ-group are equal, the elements must be conjugates.

E6: A group which is a union of finitely generated lattice-orderable groups which is not lattice-orderable

We shall define a sequence of ℓ-groups as follows. Let $G_0 = \mathbf{Z}$. Given that G_n has been defined, let G_{n+1} be the lexicographic extension $(G_n \boxplus G_n)\overleftarrow{\times}\mathbf{Z}$, where conjugation by $(0,0,1)$ interchanges components in the cardinal sum at the bottom. We now define $\alpha_n : G_n \to G_{n+1}$ by setting $\alpha_n(g) = (g,-g,0)$. It is easy to see that this is a group (but not an ℓ-group) monomorphism; we consequently may identify G_n as a subgroup of G_{n+1}, thus making

$$G = \bigcup_{n=0}^{\infty} G_n$$

a group. This example appears in Vinogradov and Vinogradov [69], who make the following observations:

a. Every element of G is conjugate to its own inverse and consequently G is not lattice-orderable; lattice-orderability is thus not a *local* property of groups.

b. Note however that the class of totally orderable groups is local (see [KK]).

E7: A group satisfying the same group sentences as a lattice-orderable group, which is not lattice-orderable

The following example is described by Vinogradov [71]. Let G be the semidirect product $\mathbf{Z}\times\mathbf{Q}$, where $(-1,0)+(0,q)+(1,0)=(0,-q)$, and then consider the groups $G_1=\mathbf{Z}\oplus G$ and $G_2=\mathbf{Z}\oplus G\oplus\mathbf{Q}$.

a. A proof similar to that in E2 shows that G_1 is not lattice-orderable.

b. We now place a lattice order on G_2 by first lattice-ordering $H=G\oplus\mathbf{Q}$ as follows: A typical element can be expressed as (m,q,r), where $m\in\mathbf{Z}$ and $q,r\in\mathbf{q}$. We then set $(m,q,r)\geq(0,0,0)$ if and only if $m>0$, or $m=0$ and $r\geq|q|$.

c. Vinogradov observes that extra copy of $\mathbf{Q}$ in G_2 does not change the first order group sentences satisfied by these groups. Thus lattice-orderable groups are not axiomatizable. This is in contrast to totally orderable groups, which are axiomatizable (by an infinite set of axioms); see Łos [54].

d. Note that this example implies that a group direct summand of a lattice-orderable group need not be lattice-orderable.

E8: Chehata's o-group

Let CH be the group of order preserving permutations of the rationals $\mathbf{Q}$ that consist of a finite number of linear pieces and have bounded support (see E53 and especially E54 for such groups with the usual lattice order). We totally order CH by calling an element positive if its rightmost nonidentity linear piece has slope less than 1. (Equivalently: if for any a in the domain of the rightmost linear piece of f we have $af>a$ then f is positive.)

a. Note that Chehata's original example [52] actually called an element positive if the leftmost nonidentity linear piece has slope greater than 1; that is, the group is "ordered from the left". Chehata also used an arbitrary ordered field in place of $\mathbf{Q}$.

b. Chehata showed that CH is algebraically simple.

c. This means that CH is a representable ℓ-group which is not an element of the torsion class $\mathcal{A}^*$.

E9: Clifford's o-group

Let CL be the group generated by $\{g_r\}$, $r\in\mathbf{Q}$, such that

$$g_r^{-1}g_sg_r=g_{\frac{(r+s)}{2}}\text{ if }s<r.$$

Then each $x\in CL$ has a unique normal form

$$x=g_{r_1}^{n_1}g_{r_2}^{n_2}g_{r_3}^{n_3}\dots g_{r_k}^{n_k}\text{ where }r_1<r_2<r_3<\dots<r_k\text{ and }n_i\neq0.$$

Order this group by calling an element positive if $n_k > 0$.

a. Clifford [52a] proved that CL is ordinally simple (but not algebraically simple).

b. Feil [80b] has shown that CL is ℓ-isomorphic to the o-group consisting of the right-bounded order preserving permutations of **Q** that consist of a finite number of linear pieces whose slopes are integral powers of $\frac{1}{2}$. In fact , if x has the normal form given above, the linear pieces break at $r_1, r_2, \ldots r_k$ and the slope of the piece with domain $[r_{i-1}, r_i]$ is $\frac{1}{2}^n$, where

$$n = \Sigma_{i=1}^{k} n_j.$$

c. In fact, if one considers the subgroup B of CL consisting of the bounded elements, the commutator subgroup of B is ordinally simple. It is unknown whether or not B is equal to its commutator subgroup (that is, B is perfect).

E10: A homogeneous distributive lattice which does not admit an ℓ-group structure

We will consider the *long line* **L**, which is constructed as follows. Let ω_1 be the first uncountable ordinal, and the set $L' = [0,1)\overleftarrow{\times}\omega_1$ ordered lexicographically from the right. Then let $\mathbf{L} = L' \backslash \{(0,0)\}$.

a. The long line is homogeneous, and since it is a totally ordered set, it is clearly distributive.

b. Since **L** is homogeneous and the cofinality and coinitiality of **L** are different and since the map $x \mapsto x^{-1}$ reverses order, **L** can not support an o-group.

Alternatively, **L** is Dedekind complete and consequently can only support only an archimedean o-group structure. However, such groups are subgroups of the reals, and hence have a countable cofinal set, which the long line lacks.

E11: A po-subgroup of an ℓ-group which is an ℓ-group but not an ℓ-subgroup

Let G be the subgroup of $C[0,1]$ consisting of the linear functions; G is clearly not an ℓ-subgroup. However, the order on $C[0,1]$ does induce an order on G which makes it an ℓ-group. For example, given two linear functions, consider the higher of the two points these functions make on the y-axis, and likewise on the line $x = 1$; the segment connecting these two points is the least upper bound in G. In fact, this description makes clear that G is ℓ-isomorphic under this order to the ℓ-group $\mathbf{R} \boxplus \mathbf{R}$. This example is due to Hofmann and appears in [C].

More generally if G is any ℓ-group (which is not totally ordered), G is a po-subgroup of F, the free ℓ-group over G as described in Chapter 6. However, Conrad [70a] has shown that in this case G is never an ℓ-subgroup of F.

E12: An ℓ-group where joins do not distribute over infinite meets in its lattice of convex ℓ-subgroups

Let G be the ℓ-group of integer-valued sequences and Σ the convex ℓ-subgroup of sequences with finite support. (That is, $G = \prod_{n\in\omega} \mathbf{Z}$ and $\Sigma = \sum_{n\in\omega} \mathbf{Z}$.) The uses to which we put this example below appear in [C].

a. For each n let $P_n = \{f \in G : f_n = 0\}$. Then

$$\Sigma \vee (\cap P_n) = \Sigma \vee \{0\} = \Sigma,$$

while

$$\bigcap(\Sigma \vee P_n) = G.$$

Thus, although the lattice of convex ℓ-subgroups is Brouwerian (see Theorem 1.2.2), the dual infinite distributive law need not hold.

b. Consider now the polars $P_n{}'$. Clearly $\vee P_n{}' = \Sigma$, which is not a polar. Thus the lattice of polars is not a complete sublattice of the lattice of convex ℓ-subgroups even though the ℓ-group in this case is strongly projectable, and so every polar is a cardinal summand.

E13: An ℓ-group whose lattice of cardinal summands is not a complete sublattice of its lattice of polars

Let H be the hyperarchimedean ℓ-group of eventually constant sequences of integers. Note that H is an ℓ-subgroup of the ℓ-group G defined in E11. For each n let

$$P_{2n} = \{g \in H : g_{2n} = 0\}.$$

Each of these is a principal polar which is clearly a cardinal summand. However

$$\bigcap P_{2n} = \{g \in H : g_i = 0, \text{ for all but a finite number of odd integers }\}.$$

This polar is not a summand, because it is evident that

$$(1,1,1,\ldots) \notin \bigcap P_{2n} \boxplus (\bigcap P_{2n})'.$$

This example appears in [C].

E14: An ℓ-group which is archimedean, special-valued and has a basis, with an ℓ-homomorphic image with none of these properties

Let H be the ℓ-group of real-valued sequences and G its convex ℓ-subgroup of sequences with finite support.

a. The ℓ-group H is archimedean, but the ℓ-group H/G isn't, because

$$n(G + (1,1,1,1,\ldots)) \leq G + (1,2,3,4,5,6,\ldots).$$

b. The ℓ-group H has a basis, but H/G doesn't, because any strictly positive element of H/G has a representation $G + \{a_n\}$, where the number of nonzero a_i's is infinite; decomposing this set into two disjoint infinite subsets gives rise to disjoint elements of H/G, both below $G + \{a_n\}$.

c. The ℓ-group H clearly has a plenary set of special values, namely the maximal primes $H_i = \{f \in H : f(i) = 0\}$. However, it is easy to see that any element in H/G has infinitely many values.

E15: A hyperarchimedean ℓ-group without a basis

Let G be the ℓ-group of periodic sequences of integers. The easiest way to see that G is hyperarchimedean is to prove that $h \in G(g) \boxplus g'$, for all $h, g \in G$, which is equivalent to being hyperarchimedean (see Conrad [74c]). It is also evident that every nonzero positive element bounds two disjoint elements, and so G does not have a basis.

E16: A divisible hyperarchimedean ℓ-group which admits no representation as a group of real-valued step functions

Let $f \in \Pi_\gamma \mathbf{R}$; we say that f is a step function if f has finite range. In [74c] Conrad considers the class of ℓ-groups which admit a representation as a group of step functions; in particular, each such ℓ-group is hyperarchimedean. The following example appears in the same paper.

Let H be the ℓ-group of real-valued sequences, and let

$$G = \{h \in H : \text{there exist } r, s \in \mathbf{Q} \text{ such that } h_i = r(\pi + \frac{1}{i}) + s,$$

$$\text{for all but a finite number of } i\}.$$

a. G is a divisible ℓ-subgroup of H which contains the small sum. Furthermore G is hyperarchimedean. However, G admits no representation as a group of step functions.

b. Now let E be the ℓ-subgroup of G consisting of the eventually constant sequences of rationals; this is a group of step functions. Furthermore, G is an a-extension of E, and so this class of ℓ-groups is not closed under a-extensions.

E17: An ℓ-group with a dense ℓ-subgroup which misses an interval sublattice

Let G be the ℓ-group of eventually constant sequences of real numbers; this is a hyperarchimedean ℓ-group with a basis. Then the Dedekind completion $G^\wedge$ of G (see Theorem 8.2.2) is the ℓ-group of bounded sequences; of course, G is a dense ℓ-subgroup of $G^\wedge$. Now,

$$a = (1, 1/2, 1/4, 1/8, \ldots) \text{ and}$$

$$b = (1/2, 1/4, 1/8, 1/16, \ldots)$$

are elements of $G^\wedge$ and evidently $a > b$. However, there exists no $c \in G$ such that $a \geq c \geq b$. This example occurs in Conrad and McAlister [69].

E18: A finitely generated archimedean ℓ-group with infinite rank as an abelian group

Let G be the ℓ-subgroup of the ℓ-group of integer-valued sequences generated by

$$(1, 1, 1, \ldots) \text{ and } (1, 2, 3, 4, \ldots).$$

It is quite easy (by taking positive parts) to show that G must contain the ℓ-subgroup of all sequences with finite support; consequently, G clearly has infinite rank as an abelian group. This example is due to Weinberg.

More generally, consider a free abelian ℓ-group on more than one generator. The Birkhoff-Baker-Beynon representation of this ℓ-group as a group of piecewise linear functions (see Chapter 6) clearly shows that this group has infinite rank.

E19: Rings of continuous functions

Let X be a topological space; then the set $C(X)$ of continuous real-valued functions on X is an archimedean ℓ-group(see 1.1.12). There is a large literature about such ℓ-groups; see in particular Anderson and Conrad [82] and Gillman and Jerison [GJ].

a. As long as the cardinality of X is nonmeasurable, $C(X)$ is laterally complete if and only if X is discrete, and in this case $C(X)$ is just the unrestricted product of $|X|$ copies of the reals (the cardinality assumption guarantees that X is real compact: see the two references mentioned above).

b. The ℓ-group $C(X)$ is complete if and only if it is strongly projectable, which is true if and only if the topological space X is extremally disconnected. Theorems along the lines of b. (and c. below) were first proved by Stone [40] and Nakano [41e].

c. The ℓ-group $C(X)$ is σ-complete (that is, every countable bounded set has a least upper bound) if and only if it is projectable, which is true if and only if the topological space is basically disconnected. Thus, an appropriate $C(X)$ serves as an example of a projectable ℓ-group which is not strongly projectable, and an ℓ-group which is σ-complete but not complete.

d. Note that ℓ-groups of the form $C(X)$ can be characterized as ℓ-groups with a collection of maximal primes whose intersection is zero, which are maximal in some sense. See Anderson and Conrad [82] and also Anderson and Blair [59].

e. Note that for any X, $C(X)$ provides an example of an archimedean a-closed ℓ-group; see Conrad [66a].

f. Consider now $X = [0, 1]$, the closed unit interval. Then $C[0, 1]$ is not projectable, from c. In fact, $C[0, 1]$ is not even semiprojectable. For consider the prime subgroups

$$P = \{f \in C[0,1] : f(\tfrac{1}{2}) = 0\},$$

$$P_l = \{f \in C[0,1] : f \equiv 0 \text{ on an interval } (a, \tfrac{1}{2}]\} \text{ and}$$

$$P_r = \{f \in C[0,1] : f \equiv 0 \text{ on an interval } [\frac{1}{2}, a)\}.$$

Then P contains both P_r and P_l but the latter two primes are incomparable.

g. Furthermore, the lattice of polar subgroups of $C[0,1]$ is not a sublattice of the lattice of all convex ℓ-subgroups. For let

$$Q = \{f \in C[0,1] : f \equiv 0 \text{ on } [1/2, 1]\}$$

and

$$Q' = \{f \in C[0,1] : f \equiv 0 \text{ on } [0, 1/2]\}.$$

Then the polar join $Q \sqcup Q' = C[0,1]$, but $Q \vee Q' = P$, the prime defined in part f.

h. Note that if G is orthofinite, then every bounded convex ℓ-subgroup has a finite basis; however, the converse is false, because $C(X)$ is clearly not orthofinite, while it has no bounded convex ℓ-subgroups. This observation occurs in [C].

i. Choose $f_{ij} \in C[0,1]^+$ so that $\int_0^1 f_{ij}(t)dt = \frac{1}{i}$ and $\bigvee_j f_{ij}$ is the constant function $\bar{1}$. We then have $\bigwedge_I \bigvee_J f_{ij} = \bar{1}$ while $\bigvee_{J^I} \bigwedge_I f_{i\alpha(i)} = \bigvee_{J^I} \bar{0} = \bar{0}$. This shows directly that $C[0,1]$ is not completely distributive. Note that in general an archimedean ℓ-group is completely distributive if and only if it has a basis (see Conrad [64]).

E20: An archimedean locally flat ℓ-group which is not hyperarchimedean

Call a prime subgroup of an ℓ-group a *minimax* prime if it is both a maximal and a minimal prime. In [80] Anderson considered ℓ-groups which have a collection of minimax primes whose intersection is zero, and called such ℓ-groups *locally flat*. If the minimax primes are all normal (which holds if the ℓ-group is normal-valued), then the ℓ-group is clearly archimedean. Since an ℓ-group is hyperarchimedean if and only if all primes are minimax (see Conrad [74c]), it follows that every hyperarchimedean ℓ-group is locally flat. Given a topological space X, call $f \in C(X)$ *locally flat* if for all $x \in X$ there exists a neighborhood U of x on which f is constant. Now let X be strongly zero dimensional: that is, the set of clopen sets forms a base for the topology; let

$$F(X) = \{f \in C(X) : f \text{ is locally flat}\}.$$

Then $F(X)$ is an archimedean locally flat ℓ-group which is a large ℓ-subgroup of $C(X)$; in fact, all archimedean locally flat vector lattices can be ℓ-embedded into some $F(X)$. Furthermore, for most X, $F(X)$ is not hyperarchimedean.

E21: Essentially closed archimedean ℓ-groups

Let X be a Stone space: that is, a compact extremally disconnected space (see the discussions in Chapter 2 and Section 8.2). Then $D(X)$, the ℓ-group of continuous almost-finite real-valued functions on X, is an essentially closed archimedean ℓ-group, and every such ℓ-group is ℓ-isomorphic to $D(X)$, for some such X (Theorem 8.2.1; Conrad [71a]).

a. Let X be a Stone space with nonmeasurable cardinality. Then the only maximal ℓ-ideals of $D(X)$ are of the form

$$D_p = \{d \in D : d(p) = 0\},$$

where p is a p-point; because of the cardinality restriction, each p-point is isolated (see [GJ] for a definition and proof). Thus if X has no isolated points, this means that $D(X)$ is an archimedean ℓ-group with no maximal ℓ-ideals, and hence in a very strong sense cannot be represented as a subdirect product of reals.

b. However, $D(X)$ is the lateral completion of $C(X)$ (since in this case $C(X)$ is complete and divisible: see E19 and Theorem 8.2.7). Thus, the lateral completion of a subdirect product of reals need not admit such a representation.

E22: A Dedekind complete, laterally complete archimedean ℓ-group which is not essentially closed

Let G be the ℓ-group of integer-valued sequences. Then G is clearly Dedekind complete and laterally complete, but it is not divisible.

a. Thus G is large in its divisible hull G^d and so is not essentially closed. Note that G^d is a proper large ℓ-subgroup of the ℓ-group H of all rational-valued sequences.

b. Since G^d contains all rational sequences with finite range, it is clear that G^d is no longer Dedekind complete.

c. Furthermore, since

$$(1/2, 1/3, 1/5, 1/7, 1/11, \ldots)$$

is not an element of G^d, it follows that G^d is not laterally complete either.

d. The essential closure of G (and of G^d) is of course the ℓ-group of all real-valued sequences.

E23: An archimedean ℓ-group which is completely distributive but not semiprojectable

Let G be the group direct sum

$$C[0,1] \oplus \sum\{\mathbf{R}_t : t \in [0,1]\}.$$

This example appears in Conrad [77]; Redfield [76] had asked whether all completely distributive archimedean ℓ-groups were semiprojectable.

a. G clearly has a basis and so is completely distributive.

b. The small sum of reals is clearly an ℓ-ideal of G and so $C[0,1]$ is an ℓ-homomorphic image of G. Now $C[0,1]$ is clearly not semiprojectable (see Example E19). Since the class $\mathcal{B}$ of semiprojectable ℓ-groups is a torsion class, this means that G is not semiprojectable either.

c. Also note that G has an ℓ-homomorphic image (namely $C[0,1]$) which is not completely distributive.

E24: An archimedean ℓ-group whose Conrad radical is not a cardinal summand

Note that for an archimedean ℓ-group, its Conrad radical (or equivalently: its distributive radical; see Corollary 10.8) consists of the intersection of polars of the form g', where g is a basic element. Thus, the Conrad radical is in this case a polar. However, it need not be a cardinal summand, as the following example (due to Conrad [64]) shows.

Consider the ℓ-group

$$H = \prod_{i=1}^{\infty} \mathbf{Z}_i \boxplus C[0,1],$$

and its convex ℓ-subgroups

$$S_1 = \sum_{i=1}^{\infty} \mathbf{Z}_i \boxplus \{0\}$$

and

$$S_2 = 0 \boxplus \{f \in C[0,1] : f(\tfrac{1}{2}) = 0\}.$$

Now let G be the subgroup of H generated by S_1, S_2 and the element g which is the constant sequence $\bar{1}$ in the first summand and the constant function $\bar{1}$ in the second summand. It is easy to check that G is in fact an ℓ-subgroup. Now the basic elements of G consist of those which live in the first summand. Thus the radical $R(G) = S_2$ while $R(G)' = S_1$. However, $g \notin S_1 \boxplus S_2$.

E25: An archimedean ℓ-group which splits but not every essential extension splits

An archimedean ℓ-group *splits* if it is a cardinal summand of each ℓ-group that contains it as a convex ℓ-subgroup. Anderson, Conrad and Kenny [77] derived a number of conditions equivalent on the one hand to splitting, and on the other to every essential extension splitting. The following example, which also appears in that paper, shows that these conditions are not equivalent.

Let H be the unrestricted direct product of copies of the reals, indexed by the closed unit interval, and let

$$G = \{g \in H : g(x) = r \text{ for some constant } r, \text{ for all but countably many } x\text{'s}\}.$$

a. G is a projectable ℓ-group containing the small sum and so its lateral completion (and hence its essential closure) is the full product H.

b. Now G does not generate H as an ℓ-ideal, which is equivalent to saying (Anderson, Conrad and Kenny [77]) that G has essential extensions which do not split.

c. However, G does split.

E26: A non-projectable archimedean ℓ-group in which each disjoint set is bounded

Recall that by Theorem 8.2.4 (due to Bernau [75b]), an archimedean laterally complete ℓ-group is necessarily projectable. The following example shows that the weaker condition that each disjoint set is bounded does not imply projectability. Note that each disjoint set being bounded in an archimedean ℓ-group is equivalent to saying that every essential extension of the ℓ-group splits (Anderson, Conrad and Kenny [77]; see example E25).

Let H be the group of all sequences of real numbers. For each $0 < h \in H$, choose $\bar{h} \geq h$ with the property that

$$\bar{h}_{2k+1} = \bar{h}_{2k+2}, \text{ for } k = 0, 1, 2, \ldots.$$

Let G be the ℓ-subgroup of H generated by $\{\bar{h} : 0 < h \in H\}$, the small sum and $g = (1, 0, 1, 0, 1, \ldots)$.

a. Since H is laterally complete and G contains elements bounding any given positive element of H, it is clear that every disjoint set of G is bounded.

b. Pick $0 < h \in H$, which is unbounded and strictly positive at each component. Then one can show that $\bar{h} \notin g'' \boxplus g'$, and hence G is not projectable.

E27: An archimedean ℓ-group which has no convex ℓ-subgroup which is maximal with respect to being a subdirect product of reals

Consider the unrestricted product of copies of $\mathbf{Z}$ indexed by the unit interval $[0, 1]$, and let K be the ℓ-group generated by the characteristic functions on closed intervals, and the set of all elements h_b, for $b \in [0, 1]$ defined as follows:

$$h_b(x) = \begin{cases} \left\lfloor \frac{1}{(x-b)^2} \right\rfloor & \text{if } x \neq b \\ 0 & \text{if } x = b, \end{cases}$$

where $\lfloor\ \rfloor$ is the greatest integer function. Then let $G = K/\Sigma$, where Σ is the small sum of integers. This example appears in Sheldon [73] (who uses it as a group of divisibility; see Chapter 11) and also in Kenny [75a].

a. Sheldon shows that G is an archimedean ℓ-group (and so the group of divisibility of a completely integrally closed domain).

b. Each element $k \in K$ can be expressed as

$$\Sigma + \sum n_i h_b.$$

Let P_b be the prime subgroup of G consisting of those elements $\Sigma + k$ for which h_b does not appear in the expression for k given above. The P_b's are exactly the maximal convex ℓ-subgroups of G, and clearly $\cap P_b$ is not zero; thus G admits no representation as a subdirect product of reals.

c. In fact, Kenny [75a] has shown that if H is a convex ℓ-subgroup of G, then H is a subdirect product of reals if and only if

$$S = \{b \in [0, 1] : P_b \cap H \subset H\}$$

has empty interior. But $[0,1]$ has no subset maximal with respect to having empty interior, and so G has no maximal convex ℓ-subgroup which can be represented as a subdirect product of reals. Thus, the class of all such ℓ-groups is not closed with respect to joins of convex ℓ-subgroups, and so is not a torsion class.

E28: An archimedean ℓ-group which admits two non-isomorphic archimedean a-closures

Note that an a-extension of an archimedean ℓ-group is archimedean (Conrad [66a]); despite this fact, a-closures of archimedean ℓ-groups need not be unique. The example of this which follows appears in the same paper.

Let H be the ℓ-group of all real-valued sequences and G its ℓ-subgroup which consists of all integer-valued sequences. We now consider the set J of all elements of H of the form $g + (r_1, r_2, r_3, \ldots)$, where $g \in G$, $0 \leq r_i < 1$, for all i, and there are only finitely many distinct r_i's. Then J is an ℓ-subgroup of H. Consider also the ℓ-subgroup K of H generated by G and the element

$$\left(\frac{\pi}{2}, \frac{3\pi}{4}, \frac{7\pi}{8}, \frac{15\pi}{16}, \ldots\right).$$

a. The ℓ-group J is an a-extension of G, and is in fact a-closed.

b. The ℓ-group K is an a-extension of G, and so there exists an a-closure of G in H which contains K. Conrad [66a] then proves that this a-closure cannot be ℓ-isomorphic to J.

E29: An abelian o-group with a nonabelian a-closure

Express the reals numbers $\mathbf{R}$ as a group direct sum $\mathbf{Q} \oplus D$, where $\mathbf{Q}$ is the group of rationals. Now consider the Hahn group $V(\Gamma, \mathbf{R})$ with root system

and let $G = \{f \in V : f(1) \in D\}$. It is obvious that V is an a-closure of G. We next define a nonabelian o-group H, with totally ordered set $\mathbf{R}\overleftarrow{\times}\mathbf{R}$ (ordered lexicographically from the right), with group operation defined as follows (where we have decomposed the second coordinates in the direct sum $\mathbf{Q} \oplus D$, so that $q, r \in \mathbf{Q}$ and $d, e \in D$):

$$(x, q + d) + (y, r + e) = (x2^r + y, q + d + r + e).$$

It is easily checked that H is also an a-extension of G. This example appears in Conrad [66a].

Note that since for o-groups the notions of a-extensions and a*-extensions coincide, this also provides an example of an abelian ℓ-group with a non abelian a*-closure.

E30: A cardinal product of Hamiltonian ℓ-groups which is not Hamiltonian

Let $K = \mathbf{Q}\overleftarrow{\times}\mathbf{Z}$, ordered lexicographically from the right, and define an addition on G as follows:

$$(q, m) + (p, n) = (2^n q + p, m + n).$$

This is a non-abelian Hamiltonian ℓ-group which is an ℓ-subgroup of the ℓ-group given in E29.

Now let G be the unrestricted cardinal product of countably many copies of K, and consider the following elements of G:

$$a = \begin{cases} 0, & 0, \quad 0, \quad 0, \ldots \\ 1, & 1, \quad 1, \quad 1, \ldots, \end{cases}$$

$$b = \begin{cases} 1, & 2, \quad 3, \quad 4, \ldots \\ 0, & 0, \quad 0, \quad 0, \ldots. \end{cases}$$

By direct computation we obtain the following conjugate:

$$a^b = \begin{cases} 0, & 0, \quad 0, \quad 0, \ldots \\ 2, & 4, \quad 8, \quad 16, \ldots. \end{cases}$$

But then $a^b \notin G(a)$, and so $G(a)$ is not normal. That is, the class of Hamiltonian ℓ-groups is not closed under direct products, and so is not a variety. This example appears in Conrad [80a].

E31: An abelian o-group which cannot be o-embedded into the Hahn group of its covering pairs

The Conrad-Harvey-Holland embedding theorem (see Theorem 3.2) asserts that every abelian ℓ-group G may be ℓ-embedded into the Hahn group $V(\Gamma(G), \mathbf{R})$. It is natural to ask whether we can replace the copies of $\mathbf{R}$ by the corresponding o-subgroups P^*/P, for each regular subgroup P. This is certainly the case if each of the o-subgroups in question is divisible (see Conrad, Harvey and Holland [63] or [C]), but the following example shows that this is false in the absence of divisibility, even in the totally ordered case. This example appears in [C].

Let H be the group $\mathbf{Q}\overleftarrow{\oplus}\mathbf{Q}$ (ordered lexicographically from the right), and let G be the subgroup generated by

$$\{\left(\frac{n}{p_n}, \frac{1}{p_n}\right) : p_n \text{ is the } n\text{th prime}\}.$$

Now G clearly has only one proper convex subgroup, which consists of those elements which live only on the bottom component, and so the root system Γ is in this case the two element chain. A straightforward argument shows that this unique proper convex subgroup is actually of the form $\{(n, 0) : n \in \mathbf{Z}\}$.

This means that the two covering pair groups consist of $\mathbf{Z}$ on the bottom and

$$\{\frac{m}{n} : n \text{ is square} - \text{free}\} \subset \mathbf{Q}$$

on top. An argument by way of contradiction then shows that G cannot be o-embedded into this Hahn group.

E32: Abelian orthofinite ℓ-groups

Let Γ be an arbitrary root system, and $V = V(\Gamma, \mathbf{R})$ the usual Hahn group. Then

$$F(\Gamma, \mathbf{R}) = \{f \in V : spt(f) \text{ lies on finitely many roots of } \Gamma\}$$

is an orthofinite ℓ-group.

Note that every abelian orthofinite ℓ-group admits a unique a-closure of the form $F(\Gamma, \mathbf{R})$ (see Conrad [66a]).

E33: An abelian finite-valued ℓ-group all of whose primes are regular, which is not orthofinite

Consider the root system Γ

and let $G = \Sigma(\Gamma, \mathbf{R})$; this example appears in Conrad [77].

a. This ℓ-group is obviously finite-valued, and the primes consist exactly of those corresponding to the elements of Γ, and hence are all regular (and special).

b. There are obviously infinite bounded disjoint sets in G and so it is not orthofinite.

E34: The Conrad radical of an ℓ-group modulo its Conrad radical need not be zero

Let Γ be the root system which consists of the real line $\mathbf{R}$ (trivially ordered) together with a copy of the real line below each point of the original copy of $\mathbf{R}$. Then let G be the ℓ-subgroup of $V(\Gamma, \mathbf{R})$, where each element is continuous on each of the real lines. This example appears in [C].

a. Note that $R(G)$ consists of all functions which are zero on the top line.

b. But $G/R(G)$ is ℓ-isomorphic to $C(\mathbf{R})$, whose Conrad radical is itself.

E35: An abelian ℓ-group which is laterally complete but not projectable

Consider the root system Γ

Denote the maximal elements of Γ by M. Let

$$G = \{f \in V(\Gamma, \mathbf{R}) : f \text{ is constant on } M\}.$$

This example appears in Conrad [77].

a. The ℓ-group G is laterally complete but not projectable.

b. The projectable hull of G is

$$G^P = \{f \in V : f \text{ has finite range on } M\}.$$

This ℓ-group is no longer laterally complete; in fact, the lateral completion of G^P is the full Hahn group V.

E36: A strongly projectable ℓ-group with a nonprojectable ℓ-homomorphic image

Let G be the ℓ-group of real-valued sequences and B the convex ℓ-subgroup of bounded sequences. Note that G is strongly projectable. Define $x, y \in G$ by setting

$$x_n = \begin{cases} p & \text{if } n \text{ is of the form } p^m, \text{ some } m, p \text{ prime} \\ 1 & \text{otherwise} \end{cases}$$

and $y_n = n$, for all n. Kenny [75a] then shows that

$$B + y \notin (B+x)'' \boxplus (B+x)'.$$

The reader should compare this to example E37; note that here the kernel B is not a closed convex ℓ-subgroup.

E37: A projectable ℓ-group with a complete ℓ-homomorphic image which is not projectable

Consider the root system Γ

and define elements $e, f \in V(\Gamma, \mathbf{Z})$ as follows:

$$e = \begin{cases} 1 \quad 0 \quad 1 \quad 0 \quad 1 \quad 0 \quad 1 \ldots \\ 1 \quad 1 \quad 1 \quad 1 \quad 1 \quad 1 \quad 1 \ldots \end{cases}$$

$$f = \begin{cases} 1 \quad 2 \quad 3 \quad 4 \quad 5 \quad 6 \quad 7 \ldots \\ 0 \quad 0 \quad 0 \quad 0 \quad 0 \quad 0 \quad 0 \ldots \end{cases}$$

Let K be the subgroup of V generated by $\Sigma(\Gamma, \mathbf{Z})$ together with e and f; then K is in fact an ℓ-subgroup. This example appears in Anderson [77].

a. Each principal polar of K has either finite or cofinite support, and so K is projectable, because $K \supseteq \Sigma(\Gamma, \mathbf{Z})$.

b. However, K is not strongly projectable, because if P is the polar corresponding to all the odd roots, then $f \notin P \boxplus P'$.

c. Let H be the convex ℓ-subgroup of K consisting of those elements which are zero on the maximal elements of Γ. Then H is closed, and K/H is ℓ-isomorphic to the ℓ-subgroup of the ℓ-group of sequences of integers generated by the small sum and the elements

$$\bar{e} = (1, 0, 1, 0, \ldots)$$

and

$$\bar{f} = (1, 2, 3, 4, \ldots).$$

This ℓ-group is not projectable because $f \notin e'' \boxplus e'$.

E38: An abelian semiprojectable ℓ-group which is not projectable

Consider the root system Γ:

Consider the following ℓ-subgroup of the Hahn group $V(\Gamma, \mathbf{R})$:

$$G = \{f \in V : f(i) = f(0), \text{ for all but a finite number of } i\}$$

This example is described in detail in Bleier and Conrad [75], Brewer, Conrad and Montgomery [74], Anderson, Conrad and Martinez [86] and Bixler and Darnel [86].

a. Let $P_n = \{f \in G : f(2) = f(3) = \ldots = f(n) = 0\}$, $n = 2, 3, \ldots$. These are all cardinal summands and principal polars. However,

$$Q = \cap P_n = \{f \in G : spt(f) = \{1\}\};$$

is a principal polar, but

$$Q' = \{f \in G : f(0) = f(1) = 0, \ spt(f) \text{ is finite}\},$$

and so the function which is 1 everywhere does not belong to $Q \boxplus Q'$. Thus G is not projectable and the lattice of cardinal summands is not a complete sublattice of the lattice $\mathcal{P}(G)$ of polars.

b. However, it is easy to check that the lattice of principal polars $\{g'' : g \in G\}$ and the lattice of coprincipal polars $\{g' : g \in G\}$ are sublattices of $\mathcal{C}(G)$.

c. The primes of G consist of $M_0 = \{f \in G : f(0) = 0\}$, $M_1 = \{f \in G : f(0) = f(1) = 0\}$, and $M_i = \{f \in G : f(i) = 0\}$ for $i = 2, 3 \ldots$. Thus each prime exceeds a unique minimal prime, and so G is semiprojectable.

d. The minimal primes are special and hence all primes of G are closed (Theorems 10.6 and Lemma 10.9). Thus $G \in \mathcal{E}ss$.

e. Any element with infinite support has infinitely many values and so G is not finite-valued, even though all convex ℓ-subgroups are closed.

f. The prime M_0 is a value for only those elements with infinite support, while $M_0 \supset M_1$, which is a special value. Thus the set of special values is not a dual ideal in $\Gamma(G)$, and so G is not special-valued, even though every element has a special value. As Bixler and Darnel [86] have pointed out, $G \in \mathcal{S}^2$, the product quasi-torsion class, because it is easy to see that M_0 is special-valued (being ℓ-isomorphic to the small sum of integers) and G/M_0 is ℓ-isomorphic to the integers.

g. The lateral completion of G is $V(\Gamma, \mathbf{R})$; however, V is not an a*-extension of G and so the a*-closure of an abelian ℓ-group need not contain its lateral completion (as was conjectured by Bleier and Conrad [73]); see the discussion of a- and a*-extensions and a- and a*-closures at the end of Chapter 3.

h. Let $H = \{f \in V : \lim_{n\to\infty} f(n) = f(0)\}$. Then H is an ℓ-subgroup of V and H is an a*-extension of G; in fact, Bleier and Conrad [75] show that H is the unique a*-closure of G. However, G is a-closed. Thus, an ℓ-group need not be a*-closed even though it is a-closed and every convex ℓ-subgroup is closed.

i. By passing to a Bezout domain with group of divisibility G, this example can be used to obtain a domain where each prime is contained in a unique maximal prime, but is not adequate; see the discussion above after Theorem 11.3 and Brewer, Conrad and Montgomery [74].

j. Consider the ℓ-subgroup $X = \{f \in G : f(1) = 0\}$, and the closed convex ℓ-subgroup P_1. Then $P_1 \cap X = P_1$ is a convex ℓ-subgroup of X; however, P_1 is not closed in X. Thus, the intersection of a closed convex ℓ-subgroup with an ℓ-subgroup need not be closed.

E39: An o-group of the form $\Sigma(\Gamma, \mathbf{R})$ which admits proper a-subspaces not ℓ-isomorphic to the original group

Let Γ be the set of all ordinal numbers less than the first uncountable ordinal ω_1, with the natural order reversed; we consider $G = \Sigma(\Gamma, \mathbf{R})$. Let

$$W = \{x \in G : \Sigma x(\gamma) = 0\}.$$

This example appears in Anderson, Bixler and Conrad [83], and is due to Fred Galvin.

a. W is an ℓ-subgroup and is a sub-vector lattice of G.

b. In fact, G is an a-extension of W. However, G is not ℓ-isomorphic to W .

E40: An abelian ℓ-group for which arbitrarily many iterations are necessary to reach its lateral completion

Let G be an ℓ-group and G^L its lateral completion. Conrad [69] first described the following transfinite process which can be used to reach G^L. Let $G(1)$ be the ℓ-subgroup of G^L generated by joins of pairwise disjoint subsets of G; inductively, define $G(\alpha+1)$ to be the ℓ-subgroup of G^L generated by joins of pairwise disjoint subsets of $G(\alpha)$, while if α is a limit ordinal, let $G(\alpha)$ be the ℓ-subgroup of G^L generated by joins of pairwise disjoint subsets of $\bigcup\{G(\beta) : \beta < \alpha\}$. A cardinality argument implies that $G^L = G(\alpha)$, for some ordinal α. If G is archimedean then $G^L = G(1)$; see Theorem 8.2.5 or Anderson, Conrad and Kenny [77].

We now define a transfinite sequence of root systems, using a double induction (with induction on all ordinals for subscripts, and on the natural numbers for superscripts). Let $\Delta_1 = \Delta_1^1$ be the root system

Having defined Δ_α^n, we obtain Δ_α^{n+1} by attaching a copy of Δ_α to each of the minimal points of Δ_α^n. Thus we have that Δ_1^2 is the root system

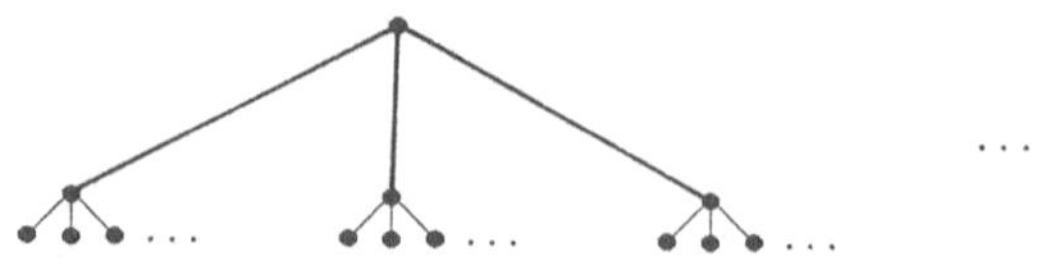

Suppose now that Δ_β has been defined, for all ordinals $\beta < \alpha$. If α is a successor ordinal, then define Δ_α by just attaching a single point above disjoint copies of each of the $\Delta^n_{\alpha-1}$'s. If α is a limit ordinal, attach a single point above copies of each of the Δ_β's , for $\beta < \alpha$. In this latter case Δ_α is the root system

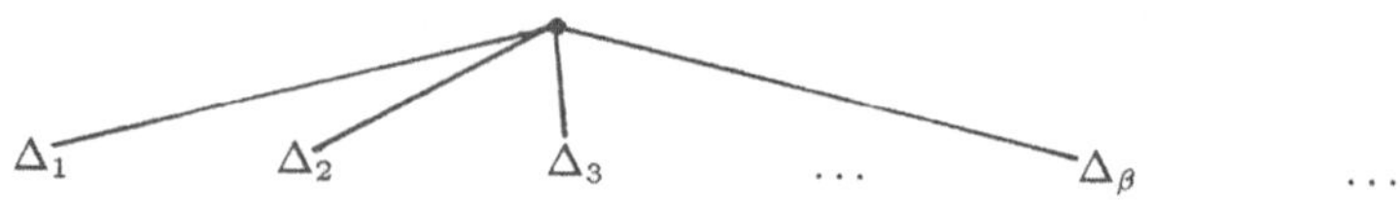

Notice then that each of the root systems has a maximum element, and every root has finite length. We will now consider the ℓ-groups $G = \Sigma(\Delta^n_\alpha, \mathbf{R})$, for all α and n.

a. Conrad [69] has shown that the lateral completion of $\Sigma(\Gamma, \mathbf{R})$ consists of the elements of $V(\Gamma, \mathbf{R})$ which have finite support on each root. Thus the lateral completion of G is the full Hahn group.

b. By means of a technical inductive argument Bixler [85] proves that $G^L = G(\alpha)$, and $G(\beta) \subset G(\alpha)$, for any ordinal $\beta < \alpha$.

E41: An ℓ-group with commuting conjugates which is not abelian

Let $G = \mathbf{Z}\overleftarrow{\times}\mathbf{Z}\overleftarrow{\times}\mathbf{Z}$, ordered lexicographically from the left, and define addition by

$$(c, b, a) + (z, y, x) = (c + z + ay, b + y, a + x).$$

This then makes G a nonabelian o-group. It is easy to check the following formula for conjugation:

$$(c, b, a)^{(z,y,x)} = (c + ay - bx, b, a).$$

This makes it easy to show that $G \in \mathcal{A}^2 \cap \mathcal{N}il$, the variety of ℓ-groups with commuting conjugates and hence is weakly abelian (see discussion of class $\mathcal{W}$). This example appears in Martinez [72b].

E42: A nonrepresentable ℓ-group with a basis

Consider $G = \mathbf{Z}$ wr $\mathbf{Z}$, the small wreath product with the usual lattice order as defined in Section 7.1.

a. Clearly G is not abelian but $G \in \mathcal{A}^2$.

b. This ℓ-group is not representable, since if we let χ be the characteristic function of $\{0\}$ and set $a = (\chi, 0)$ and $b = (0, 1)$ then $b^{-1}ab \wedge a = 1$ but $a \neq 1$.

c. It is clear that G has a basis.

d. This ℓ-group is finite-valued (and hence normal-valued) but is not orthofinite.

E43: A Hamiltonian ℓ-group which is not weakly abelian

We shall equip the small wreath product **Q** wr **Z** with a total order as follows. Given $(\langle q_n\rangle, m) \in \mathbf{Q}\ wr\ \mathbf{Z}$, consider the real number

$$p = \sum_{n=-\infty}^{\infty} q_n \pi^n,$$

which makes sense since only finitely many of the q_n's are nonzero. We then order the wreath product by setting $(\langle q_n\rangle, m)$ positive if $m > 0$ or $m = 0$ and $p > 0$. This gives us (order-theoretically) an extension of a subgroup of the reals by the integers. A direct computation verifies that this ℓ-group is not weakly abelian (see class $\mathcal{W}$). However, it is clear that every convex ℓ-subgroup is normal; that is, the ℓ-group is Hamiltonian (see class *Ham*). This example appears in Martinez [72b].

E44: An ℓ-group in $\mathcal{N}$ but not in $\bigcup \mathcal{A}^n$

Let G be the small cardinal sum $\sum \mathrm{Wr}^n \mathbf{Z}$.

a. The ℓ-group G clearly belongs to the torsion class $\mathcal{A}^*$, and is hence normal-valued.

b. However, G does not belong to any of the abelian powers $\mathcal{A}^n$. Thus, the join of a tower of varieties is not the union; recall Theorem 7.1.11.

E45: An ℓ-subgroup of a laterally complete ℓ-group that admits no lateral completion inside the ℓ-group

Let $G = \{(\langle g_n\rangle, m) \in \mathbf{Z}\ \mathrm{Wr}\ \mathbf{Z} :$ there exists $k \in \mathbf{Z}^+$ such that $g_{n_1} = g_{n_2}$ if $n_1 \equiv n_2 \pmod{k}\}$; the ℓ-group has no lateral completion inside **Z** Wr **Z**.

To demonstrate this we will give a disjoint set of elements of G such that there is no H for which $G \subseteq H \subseteq \mathbf{Z}\ \mathrm{Wr}\ \mathbf{Z}$, the join of this set is in H and G is dense in H. For each $n \in \mathbf{Z}^+$, let $x_n = (\langle g_m\rangle, 0)$ where

$$g_m = \begin{cases} 1 & \text{if } m = 2^{n-1} + 2^n.k \text{ for some } k \\ 0 & \text{otherwise.} \end{cases}$$

Note that $\{x_n\}$ is disjoint and $\bar{x} = \bigvee x_n = (\langle g_j\rangle, 0)$ where $g_j = 1$ if $j \neq 0$ and $g_0 = 0$. Let $y = (\langle y_i\rangle, 0)$ where $y_i = 1$ for all i. Then $z = y - \bar{x} = (\langle z_i\rangle, 0)$ where $z_i = 0$ for all $i \neq 0$ and $y_0 = 1$. Clearly z must be in the lateral completion of G but z bounds no element of G.

E46: A special-valued ℓ-group with a plenary set of values which cannot be used in the Ball-Conrad-Darnel embedding theorem

The following example was constructed by Bixler and Darnel [86] to show that a plenary set of a normal-valued ℓ-group must be normal in order that there exist an ℓ-embedding of the ℓ-group into a special-valued ℓ-group whose plenary set of values is isomorphic to the given plenary set (see Theorem 10.18).

Consider the splitting extension G of $C(\mathbf{R})$ by $\mathbf{R}$ defined as follows: Partially order the set $C(\mathbf{R})\overleftarrow{\times}\mathbf{R}$ by specifying that $(f, y) \geq 0$ if and only if $y > 0$ or $y = 0$ and $f \in C(\mathbf{R})^+$.

Then define addition as follows:

$$(f,a)+(g,b)=(h,a+b),$$

where $h(x)=f(x+b)+g(x)$. Consider now the following set of values for G: First, include $C(\mathbf{R})\times\{0\}$, and for each $q\in\mathbf{Q}$, include

$$M_q=\{(f,0):f(q)=0\}.$$

This set is clearly plenary, but is not normal.

If a Ball-Conrad-Darnel embedding were to exist for this plenary set, then G could be ℓ-embedded into an ℓ-group H of the form

$$\prod_{q\in\mathbf{Q}}\mathbf{R}_q\overleftarrow{\times}\mathbf{R}.$$

If this were the case, Bixler and Darnel then argue that a special element in H could have at most countably many pairwise disjoint conjugates; but a special element in H must be a component of an element from G, which can be constructed with small enough support to make this absurd.

E47: Compactly generated ℓ-groups; an essential value which is not normal in its cover

An ℓ-group is *compactly generated* if whenever a join $a=\vee A$ exists, then there is a finite subset of A of which a is the join; if this condition holds for all countable subsets, then the ℓ-group is *countably compactly generated*. These notions were introduced by Šik [62] and Jakubik [65a] respectively, and studied further by Bigard, Conrad and Wolfenstein [68]. An ℓ-group is *discrete* if each strictly positive element covers some element (here a *covers* b if $a>b$ and there exists no element c such that $a>c>b$); note in particular that a discrete ℓ-group has a basis. Bigard, Conrad and Wolfenstein proved that an ℓ-group is compactly generated if and only if it is discrete and each minimal prime is a polar. They also provide a general construction for compactly generated ℓ-groups, which involved ℓ-embedding any ℓ-group into a compactly generated one.

Let G be any ℓ-group, and let $\{M_\lambda:\lambda\in\Lambda\}$ be the set of its minimal primes; form the small sum of integers $B=\Sigma_\Lambda\mathbf{Z}$. We shall now define an ℓ-group H which is a splitting extension of B by G. Namely, we define the addition by

$$(g,b)+(h,c)=(g+h,\pi_g(b)+c),$$

where π_g permutes the coordinates of b by conjugating the corresponding minimal primes by g. We then place a lattice order on the set $G\times B$ by specifying that $(g,b)\geq 0$ if and only if $g\geq 0$ and $b_\lambda\geq 0$ whenever $g\in M_\lambda$.

What this construction accomplishes is to put a value of a basic element at the bottom of each root of the root system $\Gamma(G)$ (the reader should keep in mind that there is a one-to-one correspondence between these roots and the minimal primes). This means that H is now a compactly generated ℓ-group.

a. Note that if G is not representable (and so not all minimal primes are normal), this means that some of the essential values introduced in the ℓ-group H constructed above are not normal in their covers.

b. Bixler and Darnel [86] use a variation of this construction to obtain a proof of the Ball-Conrad-Darnel embedding theorem for normal-valued ℓ-groups (Theorem 10.18).

E48: An archimedean countably compactly generated ℓ-group which is not compactly generated

Let G be the ℓ-group of integer-valued functions on $[0,1]$ which are constant except on a finite set. Then G is clearly archimedean, countably compactly generated but not compactly generated (see E47 for definitions). Note that Bigard, Conrad and Wolfenstein [68] proved that an archimedean ℓ-group is compactly generated if and only if each ℓ-subgroup is closed, or equivalently, if and only if G is ℓ-isomorphic to a small sum of integers. This example also occurs in that paper.

E49: A discrete, finite-valued countably compactly generated ℓ-group which is not compactly generated

Let λ be an uncountable ordinal; consider the root system Γ

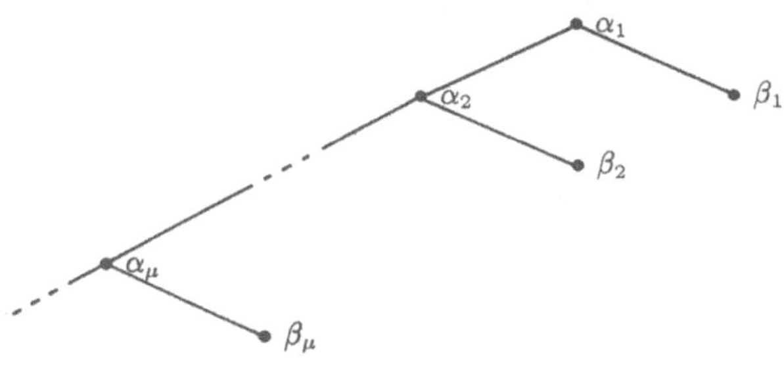

where the α's and β's are indexed by all ordinals $\mu < \lambda$. Now consider the ℓ-group $\Sigma(\Gamma, \mathbf{Z})$; this ℓ-group is clearly finite-valued, discrete and countably compactly generated, while not compactly generated (for definitions, see example A48). This example appears in Bigard, Conrad and Wolfenstein [68].

Note that an ℓ-group is compactly generated if and only if it is discrete and each minimal prime is a polar (Bigard, Conrad and Wolfenstein [68]). In this ℓ-group all minimal primes are polars of basic elements except one, and that minimal prime is actually the basic subgroup, which is clearly closed (but not a polar).

E50: Free ℓ-groups

We have of course considered free ℓ-groups in some detail in Chapter 6. We intend to emphasize here results by Kopytov [79], [83] and McCleary [85a], [85b],[85c] not proved in the text, which make free ℓ-groups useful as counterexamples. See the discussion of McCleary's methods at the end of Chapter 6. Let F_η denote the free ℓ-group on η generators.

a. The free ℓ-group F_η (with $\eta > 1$) has the following properties:

1. it is cardinally indecomposable; (actually indecomposable as a group (Glass and J. S. Wilson, unpublished))
2. it has no center; (see also Arora and McCleary [86])
3. it has no basic elements;
4. it is not completely distributive.

These also hold for free products of non-trivial ℓ-groups (see Powell and Tsinakis [84]).

b. Furthermore, McCleary [85b] has been able to give an explicit description of the root system $\mathcal{P}_\eta$ of primes of the free ℓ-group F_η. Namely, $\mathcal{P}_n$ is isomorphic to $\mathcal{P}_m$, for all integers $n, m > 1$; denote this root system by $\mathcal{P}$. Then $\mathcal{P}$ consists of 2^ω distinct copies each of the following four types of branches:

1. singletons,
2. the three element root system

3. the root system $\mathcal{P}$,
4. the root system $\mathcal{P}_\omega$ of primes of the free ℓ-group on countably many generator

These branches are then surmounted by the single point corresponding to the free ℓ-group itself. McCleary then goes on to describe the root systems $\mathcal{P}_\eta$, for larger cardinals η, in the same sort of way; the reader should consult McCleary [85b] for details.

c. Note that a free ℓ-group on finitely many generators then has uncountably many minimax primes (and in fact this holds for generator sets of all cardinalities greater than one); it is easy to conclude from McCleary's work that their intersection is zero, and so such an ℓ-group provides another example of a locally flat ℓ-group which is not normal-valued (see E54).

d. Darnel [86] and Beynon and Glass (unpublished) have described the root system of primes for free abelian ℓ-groups in a way analogous to McCleary's nonabelian work described above.

E51: A non-normal-valued ℓ-group

Consider the ℓ-group $A(\mathbf{R})$ of order-preserving permutations of the real line.

a. Holland [63] has shown that this o-2 transitive ℓ-permutation group has only three proper ℓ-ideals. These are the *right-bounded permutations*

$$R(\mathbf{R}) = \{f \in A(\mathbf{R}) : \text{there exists } r_f \in \mathbf{R} \text{ such that } y \geq r_f \text{ implies that } yf = f\};$$

the *left-bounded permutations* $L(\mathbf{R})$ (defined analogously),and the *bounded permutations*

$$B(\mathbf{R}) = R(\mathbf{R}) \cap L(\mathbf{R}).$$

b. Thus $B(\mathbf{R})$ is ordinally simple (that is, it has no proper ℓ-ideals). In fact, Lloyd [64] has shown that the above three convex ℓ-subgroups are the only normal subgroups of $A(\mathbf{R})$, and $B(\mathbf{R})$ is algebraically simple.

c. Also, $A(\mathbf{R})/R(\mathbf{R})$, $A(\mathbf{R})/L(\mathbf{R})$, $R(\mathbf{R})/B(\mathbf{R})$ and $L(\mathbf{R})/B(\mathbf{R})$ are ℓ-isomorphic simple ℓ-groups. It is interesting to note that all o-2 transitive $A(S)$ have $R(S)$, $L(S)$ and $B(S)$ as proper ℓ-ideals but this list may not be exhaustive, as E52 (due to Ball [74]) shows.

d. For $t \in \mathbf{R}$, let $M_t = \{f \in A(\mathbf{R}) : tf = t\}$, a *stabilizer subgroup* of $A(\mathbf{R})$. Then M_t is a closed maximal convex ℓ-subgroup of $A(\mathbf{R})$ (Lloyd [64]), and so is regular with cover $A(\mathbf{R})$. However from a. M_t is obviously not normal in its cover; in fact, any one stabilizer subgroup is a conjugate of any other. This shows that $A(\mathbf{R})$ is not normal-valued.

e. One can also show that $A(\mathbf{R})$ is not normal-valued as a direct consequence of the fact that any o-2 transitive ℓ-subgroup of A(S) satisfies no non-trivial equation, as was shown in the proof of Theorem 7.1.3. In fact, Ball (unpublished) has shown that $A(\mathbf{R})$ has no normal values (see [G]).

f. Note that M_t is closed and regular, but is not essential, which shows that Byrd and Lloyd's Theorem 10.6 does not hold for non-normal-valued ℓ-groups.

g. Since the stabilizer subgroups are all closed primes, the distributive radical of $A(\mathbf{R})$ is $\{1\}$; that is, $A(\mathbf{R})$ is completely distributive. However, $A(\mathbf{R})$ does not have a minimal plenary subset and does not have a basis.

h. The ℓ-group $A(\mathbf{R})$ is a-closed (Khuon [70b] and Wolfenstein [70]) and a*-closed (Glass, Holland and McCleary [75]). In fact, the same authors show that $A(\mathbf{R})$ is the unique a-closure and the unique a*-closure of $A(\mathbf{Q})$.

i. The ℓ-group $A(\mathbf{R})$ does not have unique extraction of roots. For let α be defined by $t\alpha = t + 1$ and β be defined to have period 2 with

$$t\beta = \begin{cases} \sqrt{t} + 1 & \text{if } t \in [0,1] \\ (t-1)^2 + 2 & \text{if } t \in (1,2). \end{cases}$$

Then $\alpha^2 = \beta^2$ but clearly $\alpha \neq \beta$.

E52: A totally ordered set Ω such that $A(\Omega)$ has more than $R(\Omega)$, $L(\Omega)$ and $B(\Omega)$ as proper ℓ-ideals

Let $\Omega = \mathbf{L}$ be the long line described in E10. Then $A(\mathbf{L})$ is o-2 transitive. Now let $e \leq g_0 \in A(\mathbf{L})$ be such that $(n, \alpha)g_0 = (n, \alpha)$ if and only if $0 \leq n \leq \frac{1}{2}$ and $\alpha > 0$. So $g_0 \notin R(\mathbf{L}) \cup L(\mathbf{L})$. Ball [74] has shown that the normal subgroup generated by g_0 is not $A(\mathbf{L})$, hence it is a proper normal subgroup of $A(\mathbf{L})$ other than $R(\mathbf{L})$, $L(\mathbf{L})$ and $B(\mathbf{L})$.

E53: An archimedean ℓ-group and a non-normal-valued ℓ-group with isomorphic lattices of convex ℓ-subgroups

A continuous function $f : \mathbf{R} \to \mathbf{R}$ is *finitely piecewise linear* if there exists a decomposition of $\mathbf{R}$ into finitely many intervals, on each of which f is a linear function $mx + b$, for some m and b. Consider then

$G = \{f \in C(\mathbf{R}) : f$ is finitely piecewise linear and has bounded support $\}$, and

$H = \{f \in A(\mathbf{R}) : f$ is finitely piecewise linear and has bounded support $\}$.

Roughly, H is obtained from G by rotation through $\frac{\pi}{4}$.

Ball first considered the example H. Kenoyer [84] analyzed carefully the relationship between G and H; most of the observations below appear in his paper.

a. We can completely describe the primes of G and H. Namely, for $a \in \mathbf{R}$ let

$G_a = \{f \in G : f(a) = 0\}$,

$G_a^+ = \{f \in G : f$ is 0 on some closed interval $[a, b]\}$

and $G_a^- = \{f \in G : f$ is 0 on some closed interval $[b, a]\}$.

We may similarly define subgroups H_a , H_a^+ and H_a^- of H, where elements of these subgroups are the identity permutation either at a, on an interval $[a, b]$ or on an interval $[b, a]$. In either case, each of these subgroups is prime (and in fact regular) and this is all the primes of the ℓ-group. Hence, in either case the root system of primes consists of uncountably many copies of 

b. Kenoyer [84] has shown that the isomorphism of a. between the root systems of primes can be extended to an isomorphism of the lattices $\mathcal{C}(G)$ and $\mathcal{C}(H)$ of all convex ℓ-subgroups.

c. Note that H_a is not normal in its cover H, while H_a^+ is normal in its cover H_a. Thus, H is not normal-valued; H does have a collection of normal values whose intersection is zero (equivalently: each element has a normal value), but this set is not plenary.

d. The ℓ-group G is archimedean while H is not and so being archimedean is not recognizable in the lattice of convex ℓ-subgroups. Furthermore, since G is abelian it belongs to every ℓ-group variety, while H belongs to none (except the variety of all ℓ-groups); this means that membership in any proper variety is not recognizable in the lattice of convex ℓ-subgroups.

e. Since every prime is regular, $H \in \mathcal{D}$ but is not normal-valued; it was for this purpose that Ball first considered H.

f. Since G is archimedean, all closed convex ℓ-subgroups are polars (Bigard [69a]); thus G has no closed primes, and so the closed convex ℓ-subgroups of G are not intersections of closed primes. However, the primes H_a of H are closed (McCleary [72b]), and so being a closed convex ℓ-subgroup is not recognizable in the lattice of convex ℓ-subgroups.

g. It follows immediately from the observations in part f. about closed primes that $D(G) = G$ while $D(H) = \{1\}$ (and so H is completely distributive). Thus the distributive radical is not recognizable in the lattice of convex ℓ-subgroups of an ℓ-group, and this lattice does not determine whether the ℓ-group is completely distributive.

E54: A locally flat ℓ-group which is not normal-valued

Let G be the ℓ-group of all continuous piecewise linear functions with bounded support in $A(\mathbf{Q})$ (see example E53 for definitions). Let $a \in \mathbf{R} \backslash \mathbf{Q}$; then

$$G_a = \{f \in G : f(x) = x \text{ for all } x \in (c, d) \text{ for some } c < a < d\}$$

is both a minimal and a maximal prime. Thus, G has a collection of minimax primes whose intersection is trivial, and so G is a locally flat ℓ-group (see example E20 for definitions and discussion of locally flat ℓ-groups). However, G is not normal-valued, since the primes G_α are not normal in their covers. This example should be compared to Chehata's o-group E8.

E55: Pathological ℓ-permutation groups.

An o-2 transitive ℓ-permutation group is *pathological* if it has no elements of bounded support. We give some "standard" examples of such ℓ-groups.

Let $G_1 = \{g \in A(\mathbf{R}) :$ there exists $m \in \mathbf{Z}^+$ such that for all $\alpha \in \mathbf{R}$, $(\alpha + m)g = \alpha g + m\}$.

Let $G_2 = \{h \in A(\mathbf{R}) :$ for all $\alpha \in \mathbf{R}$, there exists $m = m_h \in \mathbf{Z}^+$ such that for each $k \in \mathbf{Z}$, $(\alpha + mk)h = \alpha h + mk\}$.

Let G_3 be defined similarly to G_1 except replace $\mathbf{R}$ with $\mathbf{R}^+$. Note that the elements of G_3 with left-bounded support is an ℓ-ideal of G_3 and so G_3 is not ℓ-simple.

It is interesting to note that if G is a pathological ℓ- permutation group then $D(G) = G$ and hence G is not completely distributive. (See Corollary 8.3.5 of [G].) Note also the discussion at the end of Chapter 6 regarding McCleary's work representing free ℓ-groups pathologically.

E56: A non-normal-valued ℓ-group in which each prime subgroup contains a unique minimal prime subgroup.

Bergman (p. 219 of [G]) has constructed an example that satisfies the following lemma due to Glass and Holland (p. 218 of [G]).

Lemma Let G be an ℓ-permutation group acting o-2 transitively on Ω where each element of G has bounded support. Assume that each element of G has only finitely many bumps and that no point of $\Omega\hat{}$ (the Dedekind completion of Ω) is both the infimum of a supporting interval of an element of G and a supremum of a supporting interval of an element of G. Then each proper prime subgroup of G contains a unique minimal prime subgroup of G.

Let Ω be the set of all functions from ω_1 into $\{0,1\}$. Totally order Ω lexicographically: $\alpha < \beta$ if there exists $\mu \in \omega_1$ such that $\alpha(\mu) < \beta(\mu)$ and $\alpha(\nu) = \beta(\nu)$ if $\nu < \mu$. Ω has a largest element $\bar{1}$ and a smallest element $\bar{0}$. If $\mu < \omega_1$, p a function from μ into $\{0,1\}$ and $\alpha \in \Omega$ then $p\alpha$ is the element of Ω defined by :

$$p\alpha(\nu) = \begin{cases} p(\nu) & \text{if } \nu < \mu \\ \alpha(\lambda) & \text{if } \nu = \mu + \lambda. \end{cases}$$

A partial map f from $[p\bar{0}, p\bar{1}]$ onto $[q\bar{0}, q\bar{1}]$ is said to be *linear* if for each $\alpha \in \Omega$, $(p\alpha)f = q\alpha$.

Now let

$$G = \{g \in B(\Omega) : \text{ for some } m \in \mathbf{Z}^+ \text{ , } \Omega = \bigcup_{j=1}^{m} [p_j\bar{0}, q_j\bar{1}]$$
$$\text{and } g|[p_j\bar{0}, q_j\bar{0}] \text{ is the restriction of a linear map for } j = 1, 2, \ldots, m\}.$$

That is, G is the bounded "piecewise linear" order-preserving permutations of Ω. Let $\alpha_0 \in \Omega$ with $\alpha_0 \neq \bar{0}, \bar{1}$; then G acting on $\alpha_0 G$ is the desired ℓ-permutation group. We may fruitfully compare this example to Kenoyer's examples E53; the requirement in the lemma above on supporting intervals eliminates either the right or left hand minimal prime below each maximal prime, thus making $G \in \mathcal{B}$.

E57: The amalgamation property

A variety $\mathcal{K}$ of algebras has the *amalgamation property* if whenever G , H_1 , $H_2 \in \mathcal{K}$ and $\theta_i : G \to H_i$ are monomorphisms, $i = 1,2$, then there is an algebra L in $\mathcal{K}$ and monomorphisms $\psi_i : H_i \to L$, $i = 1,2$ such that for all $g \in G$, $g\theta_1\psi_1 = g\theta_2\psi_2$. That is, the following diagram commutes:

$$\begin{array}{ccc} G & \xrightarrow{\theta_1} & H_1 \\ \downarrow{\scriptstyle \theta_2} & & \downarrow{\scriptstyle \psi_1} \\ H_2 & \xrightarrow{\psi_2} & L \end{array}$$

We call L an *amalgam* of H_1 and H_2 over G.

Both the variety of groups and the variety of distributive lattices satisfy the amalgamation property; in fact Pierce [76] has shown that the variety $\mathcal{A}$ of abelian ℓ-groups satisfies the amalgamation property. However, the class of ℓ-groups does not as the following example due to Pierce [72] shows.

Let $H_1 = (\mathbf{Z} \boxplus \mathbf{Z})\overleftarrow{\times}\mathbf{Z}$ where $(m,n,0)$ conjugated by $(0,0,1)$ is $(n,m,0)$. Then H_1 is the Scrimger 2 (metabelian) ℓ-group. Let $a = (1,0,0)$, $b = (0,1,0)$ and $c = (0,0,1)$ and $G = (\langle a \rangle \boxplus \langle b \rangle)\overleftarrow{\times}\langle c^2 \rangle$ an abelian ℓ-subgroup of H_1. Then $A = \langle a \rangle$ and $B = \langle b \rangle$ are prime subgroups of G. Consider $\Omega = G/A \overleftarrow{\cup} G/B$, the union of G/A and G/B ordered lexicographically, and $\theta_2 : G \to A(\Omega)$ given by $(Xf)(g\theta_2) = X(fg)$ where $X = A$ or B and $f, g \in G$. Then $G\theta_2$ acting on Ω is an ℓ-permutation group.

Let $h \in A(\Omega)$ be given by $(Af)h = Af$ and $(Bf)h = Bfc^2$ where $f \in G$. Let H_2 be the ℓ-subgroup of $A(\Omega)$ generated by $G\theta_2$ and h and let $\theta_1 : G_1 \to H_1$ be the inclusion map. Pierce shows that there does not exist an ℓ-group L and monomorphisms $\psi_i : H_i \to L$ that make the above diagram commute for all $g \in G$.

The class of o-groups also does not have the amalgamation property as the following example of Glass, Saracino and Wood [84] shows. They construct an o-group G with an element $g \in G$ having no square root, then construct H_1 and H_2 containing elements h_1, h_2 respectively with $h_1^2 = g = h_2^2$ but for some $a \in G$, $h_1^{-1}ah_1 \neq h_2^{-1}ah_2$ are distinct elements of G. Hence in any group amalgam, $h_1 \neq h_2$ but $h_1^2 = h_2^2$ and so no groups amalgm is orderable.

We construct G by letting X be the rational vector space with basis $\{1, \sqrt{2}\}$ ordered as a subset of the reals. Consider the lexicographic extension $A = X\overleftarrow{\otimes}X$; note that A is a divisible abelian o-group and in fact a 4-dimensional linearly ordered rational vector space. Let $\psi = 2I_4$, an order preserving automorphism of A. Consider $\Psi = \{\psi^m : m =$

$0, \pm 1, \pm 2, \ldots\}$ ordered by $\psi^m > \psi^n$ if $m > n$. Let $G = A \overleftarrow{\rtimes} \Psi$ be a lexicographically ordered splitting extension of A by Ψ where $(a_1, \phi_1)(a_2, \phi_2) = (a_1\phi_2 + a_2, \phi_1\phi_2)$; then G is an o-group. Let

$$C = \begin{pmatrix} 0 & 1 \\ 2 & 0 \end{pmatrix}.$$

Then C is an order-preserving automorphism of X and $C^2 = 2I_2$.

Let p, q be rational numbers and

$$E_{p,q} = \begin{pmatrix} -p & q \\ -2q & p \end{pmatrix};$$

so $E_{p,q}C + CE_{p,q} = 0$. We may view

$$\theta_{p,q} = \begin{pmatrix} C & E_{p,q} \\ 0 & C \end{pmatrix}$$

as a non-singular linear transformation of A. Let $\Theta_{p,q} = \{\theta_{p,q}^m : m = 0, \pm 1, \pm 2, \ldots\}$ ordered as Ψ and let $H_{p,q} = A \overleftarrow{\rtimes} \Theta_{p,q}$ be the o-group similarly defined as $A \overleftarrow{\rtimes} \Psi$. Then ${\Theta_{p,q}}^2 = \Psi$ and the natural embedding of G in $H_{p,q}$ preserves order. If $(p_1, q_1) \neq (p_2, q_2)$ then $g = (0, \psi)$, $a = ((1,1,0,0),0)$ are the elements of G required above where $H_1 = H_{p_1,q_1}$ and $H_2 = H_{p_2,q_2}$.

The amalgamation property is closely related to the problem of embedding ℓ-groups into divisible ones. We say that a variety has the *divisible embedding property* if each member of the variety can be embedded into a divisible member of the variety. Of course, Holland's Embedding Theorem 5.4 shows that $\mathcal{L}$ enjoys this property. In group theory such embedding theorems are commonly proved using amalgamation, and in fact the amalgamation property implies the divisible embedding property. The converse is of course false, as $\mathcal{L}$ demonstrates. Unfortunately, it seems likely (see Powell and Tsinakis [85]) that no varieties of ℓ-groups except $\mathcal{A}$ satisfy the amalgamation property. Hence the divisible embedding property for varieties other than $\mathcal{L}$ and $\mathcal{A}$ (if true) must be proved by other means. In particular, the problem of embedding an o-group into a divisible o-group and the closely related ℓ-group problem of embedding a representable ℓ-group into a divisible representable one, remain unsolved. See Powell and Tsinakis [85] and [G] Chapter 10 for further discussion.

E58: Rigidly homogeneous chains. (Ohkuma groups)

If Ω is a totally ordered set such that $A(\Omega)$ is uniquely transitive (given $a, b \in \Omega$, there exists a unique $g \in A(\Omega)$ with $ag = bg$) then Ω is said to be *rigidly homogeneous.*

Ohkuma [55] has shown that rigidly homogeneous chains must be subgroups of the reals. The integers and order isomorphic sets are one such set of chains. Ohkuma has shown that there are $2^{2^{\aleph_0}}$ non-isomorphic dense subgroups of the reals that are rigidly homogeneous. A somewhat simpler proof of this can be found in Glass, Gurevich, Holland and Shelah [81] or [G], section 4.3, where the details of the following may be found.

Let p be a prime and G a torsion-free abelian group. Let $\gamma_p(G)$ be the dimension of G/pG as a vector space over $\mathbf{Z}_p$, the field of p elements. If G is rigidly homogeneous and a subgroup of the reals then $1 \leq \gamma_p(G) \leq 2^{\aleph_0}$.

Let $\{\gamma_p : p \text{ prime }\}$ be a set of cardinal numbers with $1 \leq \gamma_p \leq 2^{\aleph_0}$ and $\{I_p : p \text{ prime }\}$ a set of pairwise disjoint sets with $|I_p| = \gamma_p$. Let $I = \bigcup\{I_p : p \text{ prime }\}$ and $\{r_i : i \in I\}$ be a linearly independent set (over $\mathbf{Q}$) of reals. For each prime p, let $\mathbf{Q}_p$ be the set of rationals which, in simplest form, have denominators relatively prime to p. Then

$$\gamma_p(\mathbf{Q}_q) = \begin{cases} 0 & \text{if } q \neq p \\ 1 & \text{if } q = p \end{cases}$$

Let $G_p = \sum\{r_i Q_p : i \in I_p\}$ and $G = \sum\{G_p : p \text{ prime }\}$. Then $\gamma_p(G) = \gamma_p$ for each prime p and G is a dense subgroup of the reals. So for each prime p, there exists Ω_0, a dense subgroup of the reals with $\gamma_p(\Omega_0) = \gamma_p$.

We wish to enlarge Ω_0, preserving the invariants $\{\gamma_p : p \text{ prime }\}$ so that the only order-preserving permutations will be translations.

Let $\{f_\lambda : 1 \leq \lambda \leq 2^{\aleph_0}\}$ be an enumeration of elements of $A(\mathbf{R})$ which are not translations. (These must be killed off in the enlarged subgroup.) Let $E_0 = \{0\}$, D_0 be the divisible closure (in $\mathbf{R}$) of Ω_0. Define Ω_λ , D_λ and E_λ inductively so that

(1) D_λ is a divisible subgroup of $\mathbf{R}$ containing Ω_λ.

(2) $|D_\lambda| \leq \max\{|\lambda|, |\Omega_0|\}$.

(3) Ω_λ is a direct sum of Ω_0 and the divisible subgroup E_λ of $\mathbf{R}$.

(4) If $\lambda' < \lambda$, then $E_{\lambda'} \subseteq E_\lambda$ and $D_{\lambda'} \backslash \Omega_{\lambda'} \subseteq D_\lambda \backslash \Omega_\lambda$.

(5) $(\Omega_\lambda f_\lambda \cup \Omega_\lambda f_\lambda^{-1}) \cap (D_\lambda \backslash \Omega_\lambda) \neq 0$.

Then $\Omega = \bigcup\{\Omega_\lambda : \lambda < 2^{\aleph_0}\}$ is the required rigidly homogeneous subgroup of the reals with $\gamma_p(\Omega) = \gamma_p$.

E59: Lawless orders

Reinhold Baer asked if for every totally ordered group, the group operation can be redefined, keeping the same totally ordered set (the carrier), so that the resulting o-group is abelian. Holland, Mekler and Shelah [85] have answered this in the negative. In fact, they have constructed an o-group which is lawless: $(G, \cdot, \leq)$ is an o-group and if $(G, \star, \leq)$ is an o-group with the same carrier, then $(G, \star, \leq)$ satisfies no non-trivial equations. We outline the construction below.

For each natural number n, Let F_n be a free group on K_n, $|K_n| = n+1$ and let $H = \Pi F_n$. Fix a total order for each F_n and order H lexicographically. Then H has a countable dense subset (the restricted product) and so we may consider $(H, \leq)$ to be an ordered subset of the real line $(\mathbf{R}, \leq)$. Also, we can write elements of H as $y = (y_0, y_1, \ldots)$ where $y_n \in F_n$.

Let $\mathcal{U}$ be a non-principal ultrafilter on ω. Let $Y = \{y \in H : \{n : y_n \in K_n\} \in \mathcal{U}\}$, that is, y is "almost everywhere" a free generator on F_n. Define an equivalence relation on H with classes $[y]$ such that $[y] = [y']$ if $\{n : y_n = y'_n\} \in \mathcal{U}$. It follows that every non-trivial open interval of $(H, \leq)$ contains $2^{\aleph_0}$ pairwise disjoint non-equivalent members of Y.

Let $s = (s_0, s_1, \ldots, s_n)$ be a finite sequence with each $s_i \in F_i$, and let

$$I_s = \{y \in H : y_i = s_i, 0 \leq i \leq n\}.$$

Then I_s is an open interval of $(H \leq)$ and the collection of all such I_s forms a base for the order topology of $(H, \leq)$. Each equivalence class $[y]$ is then dense in $(H, \leq)$.

Now inductively construct G_α and R_α for each $0 \leq \alpha < 2^{\aleph_0}$ such that G_α is a subgroup of $(H, \cdot)$, R_α is a set of equivalence classes, $|G_\alpha| < 2^{\aleph_0}$, $|R_\alpha| < 2^{\aleph_0}$, and the following conditions hold:

(1) G_0 is generated by a countable dense set of pairwise inequivalent members of Y and $R_0 = \{[y] : y \in G_0 \cap Y\}$.

(2) For all $\alpha \leq \beta$, $G_\alpha \subseteq G_\beta$ and $R_\alpha \subseteq R_\beta$.

(3) For limit ordinals λ, $G_\lambda = \cup_{\alpha<\lambda} G_\alpha$ and $R_\lambda = \cup_{\alpha<\lambda} R_\alpha$.

(4) For all α, there exists $b_\alpha \in Y$ such that $G_{\alpha+1} \subseteq \langle G_\alpha, b_\alpha \rangle$ and $[b_\alpha] \in R_{\alpha+1} \backslash R_\alpha$.

(5) If α is odd, $G_{\alpha+1} = \langle G_\alpha, b_\alpha \rangle$.

(6) $G = \cup G_\alpha$.

G is the desired group. In their paper Holland, Mekler and Shelah also give a criterion for lawlessness, which G satisfies.

Bibliography

In this bibliography we have attempted to provide a comprehensive list of papers about ℓ-groups per se, especially for the last 25 years; we have by no means consulted or referred to all of the papers in this list. As is always the case in a task like this, we have used some discretion in defining the borders of our subject. We have included a few papers in such related fields as totally ordered groups, partially ordered groups, lattice-ordered rings and fields, lattice-ordered semigroups and partially ordered vector spaces. For the field of o-groups, the reader should consult the bibliographies of Mura and Rhemtulla [MR] and Kokorin and Kopytov [KK]. For more general partially ordered algebraic structures, the bibliography of Fuchs [F] is dated, but still valuable. Papers with an analytic flavor can be found in the bibliographies of Vulich [V], Luxemburg and Zaanen [LZ], Schaefer [S] and Aliprantis and Burkinshaw [AB]. We have made use of the bibliography of Bigard, Keimel and Wolfenstein [BKW] in compiling ours; their list is presently the only comprehensive one including papers on lattice-ordered rings. We have also made use of Glass's [G] invaluable annotated bibliography on permutations of a totally ordered set; the papers in the intersection of our lists consist of those which seem to have independent ℓ-group-theoretic interest.

We have first included a list of books, including volumes about partially ordered algebra, as well as a number of more general works to which we have had occasion to refer. We cite these books by means of the author(s)'s initials as listed below.

What follows next is the list of papers themselves. We cite these papers by giving the author(s)'s names, followed by a two digit citation of the year of the paper. If an author has more than one paper published in a year, we use lower case roman letters to distinguish them. Those doctoral dissertations to which we have referred are included in this list.

We have followed the conventions of the Mathematical Reviews for our bibliographic citations; we have included where possible review numbers for each item from Mathematical Reviews, in the format MR XX:YYYY, where XX is the volume number, and YYYY is the review number.

We would accept with gratitude any corrections or additions to this bibliography.

Books

ALIPRANTIS C. D; BURKINSHAW, OWEN

[AB] **Locally Solid Reisz Spaces**, Academic Press, New York, 1978; MR 58:12271.

AMAYO, RALPH, editor

[A] **Algebra, Carbondale 1980 (Proc. Conf., Southern Illinois Univ., Carbondale, Ill., 1980)**, Lecture Notes in Math., 848, Springer, Berlin, 1981; MR82d:20004.

BALL, RICHARD N.; KENNY, OTIS; SMITH, JO

[BKS] **Ordered Groups (Proc. Conf., Boise State Univ., Boise, Idaho, 1978)**, Lecture Notes in Pure and Applied Math., 62, Marcel Dekker, New York, 1980; MR81:06001.

BARWISE, JON

[Ba] **Handbook of Mathematical Logic**, North Holland, Amersterdam, 1977.

BIGARD, ALAIN; KEIMEL, KLAUS; WOLFENSTEIN, SAMUEL

[BKW] **Groupes et Anneaux Réticulés**, Springer-Verlag, Berlin, 1977; MR 58:27688, which refers to Zbl 384:06022.

BIRKHOFF, GARRETT

[B] **Lattice Theory**, 3rd. Edition, American Math. Soc. Colloq. Publ. 25, New York, 1968; earlier editions, 1940, 1948; MR 1:325, 10:673, 37:2638.

BLACK SWAMP

[BS] **The Black Swamp Problem Book**, a collection of problems in partially ordered algebra; a partial list of contents appeared in: Ordered Algebraic structures, **Notices Amer. Math. Soc.** 9 (1982), 327. Caretaker: W. Charles Holland, Bowling Green State University.

BOURBAKI, NICHOLAS

[Bo] **Algebra**, Addison-Wesley, New York, 1970.

COHN, PAUL M.

[Co] **Universal Algebra**, Harper and Row, London, 1965.

CONRAD, PAUL F.

[C] **Lattice-ordered Groups**, Tulane Lecture Notes, New Orleans, 1970.

FUCHS, LASZLO

[F] **Teilweise geordnete algebraische Strukturen**, Vandenhoeck and Ruprecht, Gottingen, 1966; MR34:4386. Review of first edition (in English) MR30:2090.

GILLMAN, LEONARD; JERISON, MEYER

[GJ] **Rings of Continuous Functions**, Van Nostrand, Princeton, 1960.

GILMER, ROBERT

[Gi] **Multiplicative Ideal Theory**, Marcel Dekker, New York, 1972.

GLASS, ANDREW M. W.

[G] **Ordered Permutation Groups**, Cambridge University Press, London, 1981; MR 83j:06004.

GLASS, ANDREW M. W.; HOLLAND, W. CHARLES

[GH] **Lattice-ordered Groups**, D. Reidel, Dordrecht, 1987.

JACOBSON, NATHAN

[Ja] **Lectures in Abstract Algebra**, Springer-Verlag, New York, 1975.

JAFFARD, PAUL

[J] **Les Systèmes d'idéaux**, Dunod, Paris, 1960; MR 22:5628.

KOKORIN, A. I.; KOPYTOV, V. M.

[KK] **Fully Ordered Groups**, Halstead Press (John Wiley and Sons), New York-Toronto, 1974; MR 51:306; Review of Russian edition MR 50:12852.

LUXEMBURG, W. A. J.; ZAANEN, A. C.

[LZ] **Reisz Spaces**, North Holland, Amsterdam, 1971.

MOČKOŘ, JIŘÍ

[M] **Groups of Divisibility**, Mathematics and its applications (East European Series), D. Reidel, Dordrecht, 1983; MR 85j:13001.

MURA, ROBERTA BOTTO; RHEMTULLA, AKBAR

[MR] **Orderable Groups**, Marcel Dekker, New York, 1977; MR 58:10652.

NAKANO, TADASI

[N] **Modern Spectral Theory**, Maruzen Co., Tokyo, 1950; MR 12:419.

NEUMANN, HANNA

[Ne] **Varieties of Groups**, Springer-Verlag, New York, 1967.

POWELL, WAYNE B.; TSINAKIS, CONSTANTINE, editors

[PT] **Ordered Algebraic Structures**, Marcel Dekker, New York, 1985.

RIBENBOIM, PAULO

[R] **Théorie des groupes ordonnés**, Monografias de matematica, Instiutio de matematica, Universidad Nacional del Sur, Bahia Blanca, 1963; MR 31:5905.

SCHAEFER, H. H.

[S] **Banach Lattices and Positive Operators**, Springer-Verlag, Berlin, 1974.

VULIKH, B. Z.

[V] **Introduction to the Theory of Partially Ordered Spaces**, Wolters-Noordhoff Scientific Publications, Groningen, 1967; MR 37:121; Review of Russian edition, MR 23:A3494.

Papers about Lattice-ordered Groups

ADYAN, S. I.

[70] Infinite irreducible systems of group identities (in Russian), **Izv. Akad. Nauk SSR. Ser. Mat.** 34 (1970), 715-734; MR 44:4078.

AMEMIYA, ICHIRÔ

[53] A general spectral theory in semiordered linear spaces, **J. Fac. Sci. Hokkaido Univ. Ser I** 12 (1953), 111-156; MR 15:137.

ANDERSON, F. W.; BLAIR, R. L.

[59] Characterizations of the algebra of all real-valued continuous functions on a completely regular space, **Illinois J. Math.** 3 (1959), 121-133; MR 20:7214.

ANDERSON, MARLOW

[77] **Subprojectable and Locally Flat Lattice-ordered Groups**, Ph.D. Dissertation, U. of Kansas, 1977.

[79] The essential closure of C(X), **Proc. Amer. Math. Soc.** 76(1979), 8-10; MR 80g:06019.

[80] Locally flat vector lattices, **Canad. J. Math.** 32 (1980), 924-936; MR 83e:06026.

[86] Hulls of Bezout domains arising from torsion-free classes of abelian ℓ-groups, to appear.

ANDERSON, MARLOW; BIXLER, PATRICK; CONRAD, PAUL

[83] Vector lattices with no proper a-subspaces, **Archiv Math. (Basel)** 41 (1983), 427-433; MR 85h:06038.

ANDERSON, MARLOW; CONRAD, PAUL

[81] Epicomplete ℓ-groups, **Algebra Universalis** 12 (1981), 224-241; MR82d:06017.

[82] The hulls of C(Y), **Rocky Mountain J. Math.** 12 (1982), 7-22; MR 84d:06026.

ANDERSON, MARLOW; CONRAD, PAUL; KENNY, OTIS

[77] Splitting properties in Archimedean ℓ-groups, **J. Austral. Math. Soc.** 23 (1977), 247-256; MR 57:202.

ANDERSON, MARLOW; CONRAD, PAUL; MARTINEZ, JORGE

[86] The lattice of convex ℓ-subgroups of a lattice-ordered group, to appear in [GH]

ANDERSON, MARLOW; EDWARDS, C.C.

[83] Unions of ℓ-groups, **Kyungpook Math. J.** 23 (1983), 91-97; MR 85a:06026.

ANDERSON, MARLOW; FEIL, TODD

[86] Pseudo-varieties of lattice-ordered groups, to appear.

ANDERSON, MARLOW; KENNY, OTIS

[80] The root system of primes of a Hahn group, **J. Austral. Math. Soc.** 29 (1980), 17-28; MR 81b:06011.

ANDERSON, MARLOW; WATKINS, JOHN

[86] Coherence of power series rings over pseudo-Bezout domains, to appear, **J. Algebra**.

ANTONOVKSII, M.JA.; MIRONOV, A.V.

[67] The theory of topological ℓ-groups (Russian), **Dokl. Akad. Nauk UzSSR** (1967), 6-8.

ARON, ELEANOR R.; HAGER, ANTHONY W.; MADDEN, JAMES J.

[82] Extension of ℓ-homomorphisms, **Rocky Mountain J. Math.** 12 (1982), 481-490; MR 83k:06019.

ARORA, A. K.

[85] Quasi-varieties of lattice-ordered groups, **Algebra Universalis** 20 (1985), 34-50.

ARORA, A. K.; MCCLEARY, STEPHEN H.

[86] Centralizers in free lattice-ordered groups, **Houston J. Math.** 12 (1986), 455-482.

BAIRD, BRIDGET B.; MARTINEZ, JORGE

[82] The automorphism group of a lattice-ordered group, **Arch. Math. (Basel)** 38 (1982), 296-310; MR 83i:06019.

[83] The semigroup of partial ℓ-isomorphisms of an abelian ℓ-group, **Acta Math. Hungar.** 41 (1983), 201-211; MR 85f:06028.

BAKER, KIRBY A.

[68] Free vector lattices, **Canad. J. Math.** 20 (1968), 58-66; MR 37:123.

BALL, RICHARD N.

[74] **Full convex ℓ-subgroups of a lattice-ordered group**, Ph.D. Dissertation, U. of Wisconsin, 1974.

[75] Full convex ℓ-subgroups and the existence of a*-closures of lattice-ordered groups, **Pacific J. Math.** 61 (1975), 7-16; MR 53:5407.

[79] Topological lattice-ordered groups, **Pacific J. Math.** 83 (1979), 1-26; MR 82h:06021.

[80a] Convergence and Cauchy structures on lattice-ordered groups, **Trans. Amer. Math. Soc.** 259 (1980), 357-392; MR 81m:06039.

[80b] Cut completions of lattice-ordered groups by Cauchy constructions, 81-92 in [BKS]; MR 82g:06028.

[81] The distinguished completion of a lattice-ordered group, 208-217 in [A]; MR 83g:06020.

[82] The generalized orthocompletion and strongly projectable hull of a lattice ordered group, **Canad. J. Math.** 34 (1982), 621-661; MR 84e:06021.

[84] Distributive Cauchy lattices, **Algebra Universalis** 18 (1984), 134-174.

[85] ℓ-group completions from lattice completions, 7-12 in [PT].

BALL, RICHARD N.; CONRAD, PAUL; DARNEL, MICHAEL

[85] Above and below subgroups of a lattice-ordered group, 13-22 in [PT].

[86] Above and below subgroups of a lattice-ordered group, **Trans. Amer. Math. Soc.**, 297 (1986), 1-40.

BALL, RICHARD N.; DAVIS, GARY

[83] The α-completion of a lattice ordered group, **Czech. Math. J.** 33(108) (1983), 111-118; MR 84g:06026.

BALL, RICHARD N.; DROSTE, MANFRED

[85] Normal subgroups of doubly transitive automorphism groups of chains, **Trans. Amer. Math. Soc.** 290 (1985), 647-664.

BANASCHEWSKI, BERNHARD

[57a] Totalgeordnete Moduln, **Arch. Math.** 7 (1957), 430-440; MR 19-385.

[57b] Uber die Vervollstandigung geordneter Gruppen, **Math. Nachr.** 16 (1957), 51-71; MR 19:388.

[64] On lattice-ordered groups, **Fund. Math.** 55 (1964), 113-122; MR 29:5930.

BARDWELL, MAUREEN A.

[79] The o-primitive components of a regular ordered permutation group, **Pacific J. Math.** 84 (1979), 261-274; MR 81f:06021.

[80a] A class of partially ordered sets whose groups of order automorphisms are lattice-ordered, 53-69 in [BKS]; MR 82c:06003.

[80b] Lattice-ordered groups of order automorphisms of partially ordered sets, **Houston Math. J.** 6 (1980), 191-225; MR 82f:06025.

BATLE, NADAL; GRANÉ, JOSÉ

[80] The Redfield topology of the lattice-ordered group of regular measures on a locally compact σ-compact topological space, **Stochastica** 4 (1980), 31-42; MR 81:22003.

BERGMAN, GEORGE

[84] Specially ordered groups, **Comm. Algebra** 12 (1984), 2315-2333.

BERMAN, JOEL

[74] Homogeneous lattices and lattice-ordered groups, **Colloq. Math.** 32 (1974), 13-24; MR 50:9726.

BERNAU, SIMON J.

[65a] On semi-normal lattice rings, **Proc. Cambridge Philos. Soc.** 61 (1965), 613-616; MR 32:1136.

[65b] Unique representation of Archimedean lattice groups and normal Archimedean lattice rings, **Proc. London Math. Soc.** 15 (1965), 599-631; MR 32:144.

[66a] Addendum to: Unique representation of Archimedean lattice groups and normal Archimedean lattice rings, **Proc. London Math. Soc.** 16 (1966), 384; MR 32:7652.

[66b] Orthocompletions of lattice groups, **Proc. London Math. Soc.** 16 (1966), 107-130; MR 32:5554.

[69] Free abelian lattice groups, **Math. Ann.** 180 (1969), 48-59; MR 39:2680.
[70] Free non-abelian lattice groups, **Math. Ann.** 186 (1970), 249-262; MR 42:160.
[72] Topologies on structure spaces of lattice groups, **Pacific J. Math.** 42 (1972), 557-568; MR 47:4889.
[74a] Hyper-archimedean vector lattices, **Indag. Math.** 36 (1974), 40-43; MR 49:5767.
[74b] Lateral completion for arbitrary lattice groups, **Bull. Amer. Math. Soc.** 80 (1974), 334-336; MR 48:10938.
[75a] The lateral completion of an arbitrary lattice group, **J. Austral. Math. Soc.** 19 (1975), 263-289; MR 52:5513.
[75b] Lateral and Dedekind completion of Archimedean lattice groups, **J. London Math. Soc.** 12 (1975/76), 320-322; MR 53:5406.
[77] Varieties of lattice groups are closed under $\mathcal{L}$-completion, **Symposia Math.** 21 (1972); MR 57:12327.
BERZ, EDGAR
[70] Uber die Distributivitat von Verbandsgruppen, **Math. Z.** 118 (1970), 191-192; MR 43:1903.
BEYNON, W.M.
[74] Combinatorial aspects of piecewise linear functions, **J. London Math. Soc.** 7 (1974), 719-727; MR 48:10943.
[75] Duality theorems for finitely generated vector lattices, **Proc. London Math. Soc.** 31 (1975), 114-128; MR 51:12655.
[77] Applications of duality in the theory of finitely generated lattice-ordered abelian groups, **Canad. J. Math.** 29 (1977), 243-254; MR 55:10350.
BIGARD, ALAIN
[66] Etude de certaines realisations des groupes réticulés, **C.R. Acad. Sci. Paris Ser. A-B** (1966), A853-A855; MR 33:2736.
[67] Propriétés des produits filtrés de groupes totalement ordonnés, **C.R. Acad. Sci. Paris Series A-B** 264 (1967), A341-A343; MR 35:1523.
[68] Sur les z-sous-groupes d'un groupe réticulé, **C.R. Acad. Sci. Paris Series A-B** 266 (1968), A261-A262; MR 38:98.
[69a] **Contribution à la théorie des groups réticulés**, Dissertation, Paris, 1969.
[69b] Groupes archimédiens et hyper-archimédiens, **Séminaire P. Dubreil, M.-L. Dubreil-Jacotin, L. Lesieur et G. Pisot: 1967/68**, Algèbre et Theorie des Nombres, Fasc. 1, Exp. 2, 13 pp. Secretariat mathematique, Paris, 1969; MR 40:4181.
[71] Sur les orthomorphisms d'un espace réticulé archimédien, **C.R. Acad. Sci. Paris Ser. A-B** 272 (1971), A10-A12; MR 43:132.
[72a] Les orthomorphisms d'un espace réticulé archimédien, **Nederl. Akad. Wetensch. Proc. Ser. A** 75 = **Indag. Math.** 34 (1972), 236-246; MR 46:7115.
[72b] Modules ordonnés injectifs, **Bull. Soc. Math. Belg.** 24 (1972), 238-248; MR 47:8390.
[73] Free lattice-ordered modules, **Pacific J. Math.** 49 (1973), 1-6; MR 49:176.
[73] Modules ordonnés injectifs, **Mathematica (Cluj)** 15(38) (1973), 15-24; MR 51:307.
BIGARD, ALAIN; CONRAD, PAUL; WOLFENSTEIN, SAMUEL
[68] Compactly generated lattice-ordered groups, **Math. Z.** 107 (1969), 201-211; MR 38:4381.
BIGARD, ALAIN; KEIMEL, KLAUS
[69] Sur les endomorphismes conservant les polaires d'un groupe réticulé archimédien, **Bull. Soc. Math. France** 97 (1969), 381-398; MR 41:6747.
BIRKHOFF, GARRETT
[42] Lattice-ordered groups, **Ann. Math.** 43 (1942), 298-331; MR 4:3.
[45] Lattice-ordered Lie groups, **Festschrift zum 60. Geburtstag von Prof. Andreas Speiser**, 209-217, Fussli, Zurich, 1945; MR 7:373.
[49] Groupes reticules, **Ann. Inst. H. Poincare** 11 (1949), 241-250; MR 11:640.
BIRKHOFF, GARRETT; PIERCE, R.S.
[56] Lattice-ordered rings, **An. Acad. Brasil. Ci.** 28 (1956), 41-69; MR 18-191.
BIXLER, PATRICK
[85] An example of the lateral completion of a lattice-ordered group, 23-32 in [PT].
(See also Anderson.)

BIXLER, PATRICK; DARNEL, MICHAEL

[86] Special-valued ℓ-groups, **Algenra Universalis** 22 (1986), 172-191.

BLEIER, ROGER D.

[71] Minimal vector lattice covers, **Bull. Austral. Math. Soc.** 5 (1971), 331-335; MR 45:5050.

[73a] Free vector lattices, **Trans. Amer. Math. Soc.** 176 (1973), 73-87; MR 47:103.

[73b] Archimedean vector lattices generated by two elements, **Proc. Amer. Math. Soc.** 39 (1973), 1-9; MR 48:8336.

[74] The SP-hull of a lattice-ordered group, **Canad. J. Math.** 26 (1974), 866-878; MR 49:10621.

[75] Free ℓ-groups and vector lattices, **J. Austral. Math. Soc.** 19 (1975), 337-342; MR 52:5518.

[76] The orthocompletion of a lattice-ordered group, **Nederl. Akad. Wetensch. Proc. Ser A79 = Indag. Math.** 38 (1976), 1-7; MR 53:L5408.

BLEIER, ROGER D.; CONRAD, PAUL

[73] The lattice of closed ideals and a*-extensions of an abelian ℓ-group, **Pacific J. Math.** 47 (1973), 329-340; MR 48:3833.

[75] a*-closures of lattice-ordered groups, **Trans. Math. Soc.** 209 (1975), 367-387;MR 53:7892.

BONDAREV, A.S.

[76] A question of P. Conrad, **Izv. Vysš. Učebn. Zaved. Mathematika** 172 (1976), 12-24; MR 58:10657.

BOSBACH, BRUNO

[82] Concerning ℓ-group (cone-) lattices, **Universal algebra (Esztergom, 1977)**, 153-159, Colloq. Math. Soc. Janos Bolyai, 29, North-Holland, Amsterdam, 1982; MR 83g:06021.

BREWER, J.W.; CONRAD, PAUL; MONTGOMERY, P.R.

[74] Lattice-ordered groups and a conjecture for adequate domains, **Proc. Amer. Math. Soc.** 43 (1974), 31-35; MR 48:10942.

BURRIS, STANLEY

A simple proof of the hereditary undecidability of the theory of lattice-ordered abelian groups, **Algebra Universalis** 20 (1985), 400-401.

BUSUEV, V.F.

[76] Topologically distributive F Ω-groups (Russian), **Modern Algebra, No. 5**, 10-19, **Leningrad Gos. Ped.Inst. im Gercena**, Leningrad, 1976; MR 58:5441.

BUSULINI, BRUNO

[58] Sulla relazione triangolare in un ℓ-gruppo, **Rend. Sem. Mat. Univ. Padova** 28 (1958), 68-70; MR 20:3918.

BUSZKOWSKI, W.

[79] Undecidability of the theory of lattice-orderable groups, **Functiones et Approximatio Comment. Math.**, 7 (1979), 23-28; MR 80:03078.

BYRD, RICHARD D.

[66] **Lattice-ordered Groups**, Ph.D. Dissertation, Tulane University, 1966.

[67] Complete distributivity in lattice-ordered groups, **Pacific J. Math.** 20 (1967), 423-432; MR 34:7680.

[68] M-polars in lattice-ordered groups, **Czech. Math. J.** 18(93) (1968), 230-239; MR 37:2651.

[69] Archimedean closures in lattice-ordered groups, **Canad. J. Math.** 21 (1969), 1004-1012; MR 39:6804.

BYRD, RICHARD D.; CONRAD, PAUL; LLOYD, JUSTIN T.

[71] Characteristic subgroups of lattice-ordered groups, **Trans. Amer. Math. Soc.** 158 (1971), 339-371; MR 43:4740.

BYRD, RICHARD D.; LLOYD, JUSTIN T.

[67] Closed subgroups and complete distributivity in lattice-ordered groups, **Math. Z.** 101 (1967), 123-130; MR 36:1371.

[69] A note on lateral completions in lattice-ordered groups, **J. London Math. Soc.** 1 (1969), 358-362; MR 40:2584.

[76] Kernels in lattice-ordered groups, **Proc. Amer. Math. Soc.** 53 (1976), 16-18; MR53:10686.

BYRD, RICHARD D., LLOYD, JUSTIN T.; MENA, R.A.; TELLER, J.R.

[77] Retractable groups, **Acta Math. Acad. Sci. Hungar.** 29 (1977), 219-233; MR 56:5386.

BYRD, RICHARD D.; LLOYD, JUSTIN T.; STEPP, JAMES W.

[77] Groups of complexes of a lattice-ordered group, **Symposia Math.** 21 (1977), 525-528; MR 57:12328.

[78] Groups of complexes of a representable lattice-ordered group, **Glasgow Math. J.** 19 (1978), 135-139; MR 58:5442.

CARTAN, HENRI

[39] Un théorème sur les groupes ordonnés, **Bull. Sci. Math.** 63 (1939), 201-205.

ČEREMISIN, A.I.

[68] Distributive F Ω-groups with a finite number of carriers (Russian), **Sibirsk. Mat. Ž.** 9 (1968), 177-187; MR 37:2652.

ČERNÁK, STEFAN

[72] Completely subdirect products of lattice-ordered groups, **Acta Fac. Rerum Natur. Univ. Comenian. Math. Publ.** 26 (1972), 121-128; MR 47:1708.

[73] The Cantor extension of a lexicographic product of ℓ-groups, **Mat. Časopis Sloven. Akad. Vied.** 23 (1973), 97-102; MR 49:2490.

[78] On some types of maximal ℓ-subgroups of a lattice-ordered group, **Math. Slovaca** 28 (1978), 349-359; MR 80h:06015.

[79] On the maximal Dedekind completion of a lattice-ordered group, **Math. Slovaca** 29 (1979), 305-313; MR 81d:06022.

CHAMBLESS, DONALD A.

[71] Representations of ℓ-groups by almost-finite quotient maps, **Proc. Amer. Math. Soc.** 28 (1971), 59-62; MR 43:131.

[72] Representation of the projectable and strongly projectable hulls of a lattice-ordered group, **Proc. Amer. Math. Soc.** 34 (1972), 346-350; MR 45:5051.

CHEHATA, C.G.

[52] An algebraically simple ordered group, **Proc. London Math. Soc.** 2 (1952), 183-197; MR 13:817.

[58] On a theorem on ordered groups, **Proc. Glasgow Math. Assoc.** 4 (1958), 16-21; MR 22:11050.

CHOE, TAE-HO

[58a] Notes on the lattice-ordered groups, **Kyungpook Math. J.** 1 (1958), 37-42; MR 20:5809; Erratum, ibid. 2 (1959), 73.

[58b] The interval topology of a lattice-ordered group, **Kyungpook Math. J.** 1 (1958), 69-74; MR 21:3492.

[60] Lattice-ordered commutative groups of the second kind, **Kyungpook Math. J.** 3 (1960), 43-48; MR 28:1240.

CLIFFORD, A.H.

[40] Partially ordered abelian groups, **Ann. Math.** 41 (1940), 465-473; MR 2:4.

[52a] A class of partially ordered abelian groups related to Ky Fan's characterizing subgroups, **Amer. J. Math.** 74 (1952), 347-356; MR 13:912.

[52b] A noncommutative ordinally simple linearly ordered group, **Proc. Amer. Math. Soc.** 2 (1952), 902-903; MR 13:625.

[54] Note on Hahn's theorem on ordered abelian groups, **Proc. Amer. Math. Soc.** 5 (1954), 860-863; MR 16:792.

CLIFFORD, A.H.; CONRAD, PAUL

[60] Lattice-ordered groups having at most two disjoint elements, **Proc. Glasgow Math. Assoc.** 4 (1960), 111-113; MR 22:9532.

COHEN, L.W.; GOFFMAN, CASPAR

[49] The topology of ordered Abelian groups, **Trans. Amer. Math. Soc.** 67 (1949), 310-319; MR 11:324.

COHN, P. M.

[57] Groups of order automorphisms of ordered sets, **Mathematika** 4 (1957), 41-50; MR 19:940.

CONRAD, PAUL

[53] Embedding theorems for abelian groups with valuations, **Amer. J. Math.** 75 (1953), 1-29; MR 14:842.

[54] On ordered division rings, **Proc. Amer. Math. Soc.** 5 (1954), 323-328; MR 15:849.

[55] Extensions of ordered groups, **Proc. Amer. Math. Soc.** 6 (1955), 516-528; MR17:458.

[58a] The group of order preserving automorphisms of an ordered abelian group, **Proc. Amer. Math. Soc.** 9 (1958), 382-389; MR 21:1340.

[58b] A note on valued linear spaces, **Proc. Amer. Math. Soc.** 9 (1958), 646-647; MR 20:5808.

[58c] On ordered vector spaces, **J. Indian Math. Soc.** 22 (1958), 27-32; MR 21:4983.

[59a] Right ordered groups, **Michigan Math. J.** 6 (1959), 267-275; MR 21:5684.

[59b] A correction and an improvement of a theorem on ordered groups, **Proc. Amer. Math. Soc.** 10 (1959), 182-184; MR 21:3490.

[59c] Non-abelian ordered groups, **Pacific J. Math.** 9 (1959), 25-41; MR 21:3491.

[60] The structure of a lattice-ordered group with a finite number of disjoint elements, **Michigan Math. J.** 7 (1960), 171-180; MR 22:6854.

[61] Some structure theorems for lattice-ordered groups, **Trans. Amer. Math. Soc.** 99 (1961), 212-240; MR 22:12143.

[62] Regularly ordered groups, **Proc. Amer. Math. Soc.** 13 (1962), 726-731; MR 26:3794.

[64] The relationship between the radical of a lattice-ordered group and complete distributivity, **Pacific. J. Math.** 14 (1964), 493-499; MR 29:3556.

[65] The lattice of all convex ℓ-subgroups of a lattice-ordered group, **Czech. Math. J.** 15(90) (1965), 101-123; MR 30:3926.

[66a] Archimedean extensions of lattice-ordered groups, **J. Indian Math. Soc.** 30 (1966), 131-160; MR 37:118.

[66b] Representation of partially ordered abelian groups as groups of real valued functions, **Acta Math.** 116 (1966), 199-221; MR 34:1418.

[67a] **Introduction à la théorie des groupes réticulés**, Cours rédigé par Bernard Gostiaux et Samuel Wolfenstein, Université de Paris, Faculté des Sciences de Paris, Sécrétariat mathématique, Paris, 1967; MR 37:1289.

[67b] A characterization of lattice-ordered groups by their convex ℓ-subgroups, **J. Austral. Math. Soc.** 7 (1967), 145-159; MR 35:5371.

[67c] Lateral completions of lattice-ordered groups, **Proc. Internat. Conf. Theory of Groups, Australian Nat. U., Canberra**, Gordon and Breach, 1967.

[68a] Lex-subgroups of lattice-ordered groups, **Czech. Math. J.** 18(93) (1968), 86-103; MR 37:1290.

[68b] Subdirect sums of integers and reals, **Proc. Amer. Math. Soc.** 19 (1968), 1176-1182; MR 38:1044.

[68c] Lifting disjoint sets in vector lattices, **Canad. J. Math.** 20 (1968), 1362-1364; MR 38:1043.

[69] The lateral completion of a lattice-ordered group, **Proc. London Math. Soc.** 3(19) (1969), 444-480; MR 39:5442.

[70a] Free lattice-ordered groups, **J. Algebra** 16 (1970), 191-203; MR 42:5875.

[71a] The essential closure of an Archimedean lattice-ordered group, **Duke Math. J.** 38 (1971), 151-160; MR 43:3190.

[71b] Free abelian ℓ-groups and vector lattices, **Math. Ann.** 190 (1971), 306-312; MR 43:7382.

[71c] Minimal vector lattice covers, **Bull. Austral. Math. Soc.** 4 (1971), 35-39; MR 42:7573.

[73] The hulls of representable ℓ-groups and f-rings, Collection of articles dedicated to the memory of Hanna Neumann, IV, **J. Austral. Math. Soc.** 16 (1973), 385-415; MR 49:8913.

[74a] Countable vector lattices, **Bull. Austral. Math. Soc.** 10 (1974), 371-376; MR 52:13563.

[74b] The additive group of an f-ring, **Canad. J. Math.** 26 (1974), 1157-1168; MR 50:6965.

[74c] Epi-archimedean groups, **Czech. Math. J.** 24(99) (1974), 192-218; MR 50:203.

[74d] The topological completion and the linearly compact hull of an abelian ℓ-group, **Proc. London Math. Soc.** (3)28 (1974), 457-482; MR 49:4899.

[75a] Changing the scalar multiplication on a vector lattice, **J. Austral. Math. Soc.** 20 (1975), 332-347; MR 52 7998.

[75b] The hulls of semiprime rings, **Bull. Austral. Math. Soc.** 12 (1975), 311-314.

[76] Review of Bernau [75a], MR52:5513.

[77] Torsion radicals of lattice-ordered groups, **Symposia Math.** 21 (1977), 479-513; MR 57:5855.

[78] The hulls of semiprime rings, **Czech. Math. J.** 28 (1978), 59-86; mR 57:3179.

[80a] Minimal prime subgroups of lattice-ordered groups, **Czech. Math. J.** 30(105) (1980), 280-295; MR 81d:06023.

[80b] The structure of an ℓ-group that is determined by its minimal prime subgroups, 1-27 in [BKS]; MR 82b:06015.

[81] K-radical classes of lattice-ordered groups, 186-207 in [A]; MR 82h:06022.

(See also Anderson, Ball, Bigard, Bleier, Brewer, Byrd and Clifford.)

CONRAD, PAUL; DARNEL, MICHAEL

[86] ℓ-groups with a unique addition, **Proc. of 1st International Symposium on Ordered Algebra, Luminy, France**, 1984.

CONRAD, PAUL; DAUNS, JOHN

[69] An embedding theorem for lattice ordered fields, **Pacific J. Math.** 30 (1969), 385-398; MR 40:128.

CONRAD, PAUL; DIEM, J.E.

[71] The ring of polar preserving endomorphisms of an abelian lattice-ordered group, **Illinois J. Math.** 15 (1971), 222-240; MR 44:2680.

CONRAD, PAUL; HARVEY, JOHN; HOLLAND, W. CHARLES

[63] The Hahn embedding theorem for lattice-ordered groups, **Trans. Amer. Math. Soc.** 108 (1963), 143-169; MR 27:1519.

CONRAD, PAUL; MCALISTER, DONALD

[69] The completion of a lattice-ordered group, **J. Austral. Math. Soc.** 9 (1969), 182-208; MR 40:2585.

CONRAD, PAUL; MCCARTHY, PAUL

[73] The structure of f-algebras, **Math. Nachr.** 58 (1973), 169-191; MR 48:8339.

CONRAD, PAUL; MONTGOMERY, PHILIP

[75] Lattice-ordered groups with rank one components, **Czech. Math. J.** 25(100), 445-453; MR 52:7993.

CONRAD, PAUL; TELLER, J. ROGER

[70] Abelian pseudo lattice ordered groups, **Publ. Math. Debrecen** 17 (1970), 223-241; MR48:2017.

CORNISH, WILLIAM H.

[80] Lattice-ordered groups and BCK algebras, **Math. Japon.** 25 (1980), 471-476; MR 81m:06040.

DARNEL, MICHAEL

[86a] Closure operators on radical classes of lattice-ordered groups, **Czech. Math. J.** 37(112) (1987), 51-64.

[86b] The structure of free abelian ℓ-groups and free vector lattices, to appear.

[86c] Special-valued ℓ-groups and abelian covers, to appear.

(See also Ball, Bixler and Conrad.)

DASHIELL, F.; HAGER, ANTHONY; HENDRIKSEN, MELVIN

[80] Order Cauchy completions of rings and vector lattices of continuous functions, **Canad. J. Math.** 32 (1980), 657-685; MR 81k:46020.

DAUNS, JOHN

[69] Representation of L-groups and F-rings, **Pacific J. Math.** 31 (1969), 629-654; MR 41:130.

(See also Conrad.)

DAVIS, GARY

[75a] Orthogonality relations on abelian groups, **J. Austral. Math. Soc.** 19 (1975), 173-179; MR 51:12654.

[75b] Compatible tight Riesz orders on groups of integer-valued functions, **Bull. Austral. Math. Soc.** 12 (1975), 383-390; MR 54:5075.

[80] Permutation representation of groups with Boolean orthogonalities, **J. Austral. Math. Soc.** Ser. A 30 (1980/81), 412-418; MR 82h:06023.

(See also Ball.)

DAVIS, GARY; FOX, COLIN D.

[76] Compatible tight Riesz orders on the group of automorphisms of an o-2 transitive set, **Canad. J. Math.** 28 (1976), 1076-1081; MR 54:7350; addendum, **ibid.**, 29 (1977), 664-665; MR 55:7879.

DAVIS, GARY; LOCI, ELIZABETH

[76] Compatible tight Riesz orders on ordered permutation groups, **J. Austral. Math. Soc.** 21 (1976), 317-333; MR 53:13068.

DAVIS, GARY; MCCLEARY, STEPHEN H.

[81a] The lateral completion of a completely distributive lattice-ordered group (revisited), **J. Austral. Math. Soc.** 31 (1981), 114-128; MR 82k:06021.

[81b] The π-full tight Riesz orders on $A(\Omega)$, **Canad. Math. Bull.** 24 (1981), 137-151; MR 83b:06015.

DEDEKIND, RICHARD

[97] Über Zerlungen von Zahlen durch ihre größten gemeinsamen Teiler, 103-148 in **Ges. Werke**, Vol. 2, Braunshweig, 1931.

DIEUDONNE, JEAN

[41] Sur la théorie de la divisibilité, **Bull. Soc. Math. France** 69 (1941), 133-144; MR 7:110.

DLAB, V.

[68] On a family of simple ordered groups, **J. Austral. Math. Soc.** 8 (1968), 591-608; MR 37:3978.

DOMRAČEVA, G.I.

[69] Extension of ideals in lattice-ordered algebras (in Russian), **Leningrad. Gos. Ped. Inst. Učen. Zap. 440 Nekotorye Voprosy Vysš. Mat. i Metodiki i Prepodav. Mat.** (1969), 3-11; MR 50:6966.

DREVENJAK, ONDREI

[75] Lattice-ordered distributive Ω-groups with a basis (in Russian), **Mat. Časopis Sloven. Akad. Vied** 25 (1975), 11-21; MR 52:3003.

[80] Lexicographic Ω-product of lattice-ordered F Ω-groups (in Russian), **Math. Slovaca** 30 (1980), 31-50; MR 81f:06022.

DREVENJAK, ONDREI; JAKUBIK, JAN

[72] Lattice-ordered groups with a basis, **Math. Nachr.** 53 (1972), 217-236; MR 48:2018.

DROSTE, MANFRED; SHELAH, SAHARON

[85] A construction of all normal subgroup lattices of 2- transitive automorphism groups of linearly ordered sets, **Israel J. Math.** 51 (1985), 223-261.

EISENBUD, DAVID

[69] Groups of order automorphisms of certain homogeneous ordered sets, **Michigan Math. J.** 16 (1969), 59-63; MR 39:1373.

ELLIOTT, GEORGE A.

[78] A property of totally ordered abelian groups, **C.R. Math. Rep. Acad. Sci. Canada** 1 (1978/79), 63-66; MR 80c:06023.

[79] On totally ordered groups, and K_0, **Ring Theory (Proc. Conf., Univ. Waterloo, Waterloo, 1978)**, 1-49, Lecture Notes in Math., 734, Spring, Berlin, 1979; MR 81g:06012.

ELLIS, JOHN

[68] **Group topological convergence in completely distributive lattice-ordered groups**, Ph.D. Dissertation, Tulane, New Orleans, 1968.

EVERETT, C.J.

[44] Sequence completion of lattice moduls, **Duke Math. J.** 11 (1944), 109-119; MR 5:169.

[50] Note on a result of L. Fuchs on ordered groups, **Amer. J. Math.** 72 (1950), 216; MR 11:324.

EVERETT, C.J.; ULAM, STANISLAV

[45] On ordered groups, **Trans. Amer. Math. Soc.** 57 (1945), 208-216; MR 7:4.

FEIL, TODD

[80a] **Varieties of Representable Lattice-ordered Groups**, Ph.D. Dissertation, Bowling Green State U., 1980.

[80b] A comparison of Chehata's and Clifford's ordinally simple ordered groups, **Proc. Amer. Math. Soc.** 79 (1980), 512-514; MR 81h:06014.

[82] An uncountable tower of ℓ-group varieties, **Algebra Universalis** 14 (1982), 129-131; MR 83d:06022.

[85] Varieties of representable ℓ-groups, 77-88 in [PT].

(See also Anderson.)

FIALA, FRANTIŠEK

[67a] Über einen gewissen Ultraantifilterraum, **Math. Nachr.** 33 (1967), 231-249; MR 36:88.

[67b] Verbandsgruppen mit o-kompakten Komponentenverbanden, **Arch. Math. (Brno)** 3 (1967), 177-184; MR 39:1374.

[68] Standard-Ultraantifilter im Verband aller Komponenten einer ℓ-Gruppe, **Acta Math. Acad. Sci. Hungar.** 19 (1968), 405-412; MR 38:430.

FLEISCHER, ISIDORE

[56] Functional representation of partially ordered groups, **Ann. Math.** (2)64 (1956), 260-263; MR 18:136.

[72] Remarks on real representations and comments on Conrad's paper, **Norske Vid. Selsk. Skr. (Trondheim) (1972)**, no. 10, 4 pp.; MR 46:7114.

[73] A final remark on extending to strict total orders in modules, **Bull. Austral. Math. Soc.** 9 (1973), 137-140; MR 49:177.

[76] Über verbandsgeordnete Vektorgruppen mit Operatoren, **Math. Nachr.** 72 (1976), 141-144; MR 54:200.

[77a] Remarks on "Embedding theorems and generalized discrete ordered abelian groups" (**Trans. Amer. Math. Soc.** 175 (1973), 283-297) by P. Hill and J.L. Mott, **Trans. Amer. Math. Soc.** 231 (1977), 273-274; MR 56:2896.

[77b] The automorphism group of a scattered set can be non-commutative, **Bull. Austral. Math. Soc.** 16 (1977), 306; MR 57:6157.

[81] The Hahn embedding theorem: analysis, refinements, proof, 278-290 in [A]; MR83b:06016.

FOX, COLIN D.

[83] On the Scrimger varieties of lattice-ordered groups, **Algebra Universalis** 16 (1983), 163-166; MR 85e:06022.

(See also Davis.)

FRANCHELLO, JAMES D.

[78] Sublattices of free products of lattice-ordered groups, **Algebra Universalis** 8 (1978), 101-110; MR 56:8461.

FREUDENTHAL, H.

[36] Teilweise geordnete Moduln, **Nederl. Akad. Wetensh. Proc.** 39 (1936), 641-651.

FRIED, E.

[65] Representation of partially ordered groups, **Acta Sci. Math. (Szeged)** 26 (1965), 15-18; MR 31:1307.

[67] Note on the representation of partially ordered groups, **Ann. Univ. Sci. Budapest Eotvos Sect. Math.** 10 (1967), 85-87; MR 39:1375.

FUCHS, LÁSZLÓ

[49] Absolutes in partially ordered groups, **Nederl. Akad, Wetensch., Proc.** 52 (1949), 251-255 = **Indag. Math.** 11 (1949), 66-70; MR 11:9.

[50a] On partially ordered groups, **Nederl. Akad. Wetensch., Proc.** 53 (1950), 828-834 = **Indag. Math.** 12 (1950), 272-278; MR 12:10.

[50b] The extension of partially ordered groups, **Acta Math. Acad. Sci. Hungar.** 1 (1950), 118-124; MR 13:436.

[50c] On the extension of the partial order of groups, **Amer. J. Math.** 72 (1950), 191-194; MR 11:323.

[59] Notes on ordered groups and rings, **Fund. Math.** 46 (1959), 167-174; MR 20:7069.

[65a] Approximation of lattice-ordered groups, **Ann. Univ. Sci. Budapest. Eotvos Sect. Math.** 8 (1965), 187-203; MR 41:126.

[65b] On partially ordered algebras. II., **Acta.Sci. Math. (Szeged)** 26 (1965), 35-41; MR 31:4749.

[65c] Riesz groups, **Ann. Scuola Norm. Sup. Pisa** (3)19 (1965), 1-34; MR 31:4843.

[65d] Note on the class of partially ordered groups, **Bull. Acad. Polon. Ser. Sci. Math. Astronom. Phys.** 13 (1965), 757-759; MR 33:203.

FUKAMIYA, MASANORI; YOSIDA, KOSAKU

[41] On vector lattice with a unit. II., **Proc. Imp. Acad. Tokyo** 17 (1941), 479-482; MR 7:409.

GAVALČOVÁ, TAT'JANA

[72] Decomposition of a complete L-group into an M-product of an M-atomic and an M-nonatomic L-subgroup, **Acta Fac. Rerum Natur. Univ. Comenian. Math. Publ.** 28 (1972), 91-97; MR 47: 3272.

[74] σ-compactness in ℓ-groups, **Mat. Časopis Sloven. Akad. Vied** 24 (1974), 21-30; MR 50:9739.

GEILER, V. A.; VEKSLER, A. I.

[72] Order and disjoint completeness of linear partially ordered spaces, **Sibirsk. Math. Z.** 13 (1972), 43-51; MR 45:5713.

GLASS, ANDREW M. W.

[72a] Which abelian groups can support a directed, interpolation order?, **Proc. Amer. Math. Soc.** 31 (1972), 395-400; MR 44:6580.

[72b] An application of ultraproducts to lattice-ordered groups, **Canad. J. Math.** 24 (1972), 1063-1064; MR 47:96.

[72c] Polars and their applications in directed interpolation groups, **Trans. Amer. Math. Soc.** 166 (1972), 1-25; MR 45:5052.

[73] Archimedean extensions of directed interpolation groups, **Pacific J. Math.** 44 (1973), 515-521; MR 48:202.

[74a] The word problem for lattice-ordered groups, **Proc. Edinburgh Math. Soc.** (2)19 (1974/75), 217-219; MR 51:3313.

[74b] ℓ-simple lattice-ordered groups, **Proc. Edinburgh Math. Soc.** (2)19 (1974/75), 133-138; MR 53:13069.

[75] Results in partially ordered groups, **Comm. Algebra** 3 (1975), 749-761; MR 52:13558.

[76a] **Ordered Permutation Groups**, Bowling Green State University, Bowling Green, Ohio, 1976; MR 54:10097.

[76b] Compatible tight Riesz orders, **Canad. J. Math.** 28 (1976), 186-200; MR 53:13070.

[79] Compatible tight Riesz orders. II., **Canad. J. Math.** 31 (1979), 304-307; MR 80h:06014.

[81b] Elementary types of automorphisms of linearly ordered sets - a survey, 218-229 in [A]; MR 82e:06001.

[83] Countable lattice-ordered groups, **Math. Proc. Cambridge Philos. Soc.** 94 (1983), 29-33; MR 84k:06019.

[84a] The isomorphism problem and undecidable properties for finitely presented lattice-ordered groups, 152-170 in **Orders: description and roles (L'Arbresle, 1982)**, North Holland, 1984; MR 86m:06030.

[84b] A directed d-group that is not a group of divisibility, **Czech. Math. J.**, 34(109) (1984), 475-476.

[85] Effective extensions of lattice-ordered groups that preserve the degree of the conjugacy and the word problem, 89-98 in [PT].

[86] Generating varieties of lattice-ordered groups; approximating wreath products, **Illinois J. Math.** 30 (1986), 214-221.

GLASS, ANDREW M. W.; GUREVICH, YURI

[83] The word problem for lattice-ordered groups, **Trans. Amer. Math. Soc.** 280 (1983), 127-138; MR 85d:06015.

GLASS, ANDREW M.W.;GUREVICH, YURI; HOLLAND, W. CHARLES; JAMBU-GIRAUDET, M.

[81] Elementary theory of automorphism groups of doubly homogeneous chains, **Logic Year 1979-1980: Proc. Sem. and Conf. Math. Logic, U. Conn. Storrs**, 1981; MR 82i:03047.

GLASS, ANDREW M. W.; GUREVICH, YURI; HOLLAND, W. CHARLES; SHELAH, SAHARON

[81] Rigid homogeneous chains, **Math. Proc. Cambridge Philos. Soc.** 89 (1981), 7-17; MR 82c:06001.

GLASS, ANDREW M. W.; HOLLAND, W. CHARLES

[73] A characterization of normal-valued lattice-ordered groups, **Notices Amer. Math. Soc.** 20 (1973), #73T-A237, p. A563.

GLASS, ANDREW M. W.; HOLLAND, W. CHARLES; MCCLEARY, STEPHEN H.

[75] a*-closures of completely distributive lattice-ordered groups, **Pacific J. Math.** 59 (1975), 43-67; correction: ibid., 61 (1975), 606; MR 52:7994.

[80] The structure of ℓ-group varieties, **Algebra Universalis** 10 (1980), 1-20; MR81k:06026a.

GLASS, ANDREW M. W.; MADDEN, JAMES L.

[84] The word problem versus the isomorphism problem, **J. London Math. Soc.** (2) 30, (1984), 53-61.

GLASS, ANDREW M. W.; MCCLEARY, STEPHEN H.

[76] Some ℓ-simple pathological lattice-ordered groups, **Proc. Amer. Math. Soc.** 57 (1976), 221-226; MR 53:7894.

GLASS, ANDREW M. W.; PIERCE, KEITH R.

[80a] Existentially complete abelian lattice-ordered groups, **Trans. Amer. Math. Soc.** 261 (1980), 255-270; MR 81k:03032.

[80b] Existentially complete lattice-ordered groups, **Israel J. Math.** 36 (1980), 257-272; MR 82b:06016.

[80c] Equations and inequations in lattice-ordered groups, 141-171 in [BKS]; MR 82c:06030.

GLASS, ANDREW M. W.; SARACINO, DAN; WOOD, CAROL

[84] Nonamalgamation of ordered groups, **Math. Proc. Cambridge Philos. Soc.** 95 (1984), 191-195 MR 86c:06020.

GOFFMAN, CASPAR

[57] A lattice homomorphism of a lattice ordered group, **Proc. Amer. Math. Soc.** 8 (1957), 547-550; MR 19:388.

[58a] Remarks on lattice ordered groups and vector lattices I. Caratheodory functions, **Trans. Amer. Math. Soc.** 88 (1958), 107-120; MR 20:3800.

[58b] A class of lattice-ordered algebras, **Bull. Amer. Math. Soc.** 64 (1958), 170-173; MR 20:3801.

[59] Completeness in topological vector lattices, **Amer. Math. Monthly** 66 (1959), 87-92; MR 20:6023.

(See also Cohen.)

GOODEARL, K.R.; HANDELMAN, D.E.

[80] Metric completions of partially ordered abelian groups, **Indiana Univ. Math. J.** 29 (1980), 861-895; MR 82b:06020.

GRANÉ MANLLEU, JOSÉ

[80a] On the surjectivity of the Ribenboim representation of a lattice-ordered group, **Stochastica** 4 (1980), 89-90; MR 81j:06021.

[80b] On the isometries of lattice-ordered groups and rings, **Stochastica** 4 (1980), 103-127; MR 82h:06027.

[81a] Topological properties of the groups of isometries of a lattice-ordered group, **Stochastica** 5 (1981), 89-94; MR 82m:06018.

[81b] Invariant sets for the group of isometries of a lattice-ordered group, **Stochastica** 5, 163-167; MR 83i:06020.

[83] Continuity of the isometries of a lattice-ordered group in the Redfield topology, 59-63 in **Proc. of the 12th annual Conf. of Spanish Math. (Malaga, 1976)**, Malaga, 1983.

(See also Batle.)

GRAVETT, K.A.H.

[55] Valued linear spaces, **Quart. J. Math. Oxford** (2) 6 (1955), 309-315; MR 19:385.

[56] Ordered abelian groups, **Quart. J. Math. Oxford** (2) 7 (1956), 57-63; MR 19:1037.

GURCHENKOV, S.A.

[82a] Minimal varieties of ℓ-groups, **Algebra i Logika** 21 (1982), 131-137; MR 85a:06027.

[82b] Varieties of nilpotent lattice-ordered groups (Russian), **Algebra i Logika** 21 (1982), 499-510; MR 85d:06016.

[84a] Coverings in the lattice of ℓ-varieties (Russian), **Mat. Zametki** 35 (1984), 677-684; MR 86f:06026.

[84b] Varieties of ℓ-groups with the identity $x^p y^p = y^p x^p$ have finite basis, **Algebra and Logic** 23 (1984), 20-35; MR 86m:06031.

[85] Varieties of ℓ-groups with infinite axiomatic rank (Russian), **Sibirsk. Mat. Z.** 26 (1985), 66-70, 225; MR 87b:06029.

GURCHENKOV, S.A.; KOPYTOV, V.M.

[87] On covers of the variety of abelian lattice-ordered groups (Russian), **Siberian Math J.** 28 (1987).

GUREVICH, YURI

[67a] Hereditary undecidability of a class of lattice-ordered Abelian groups (in Russian), **Algebra i Logika** 6 (1967), 45-62; MR 36:92.

[67b] On the elementary theory of lattice-ordered abelian groups and K-lineals (in Russian), **Dokl. Akad. Nauk SSSR** 175 (1967), 1213-1215; MR 36:1373.

GUREVICH, YURI; HOLLAND, W. CHARLES

[81] Recognizing the real line, **Trans. Amer. Math. Soc.** 265 (1981), 527-534.

GŬREVICH, E.E.

[68] Convergence with a regulator in commutative ℓ-groups (in Russian), **Mat. Zametki** 3 (1968), 279-284; MR 37:3979.

GURVIČ-LEĬNOV, A.H.; RABINOVIČ, E.B.

[77] The representation of all automorphisms of a homogeneous linear order by a wreath product of o-primitive components (in Russian), **Vestnik Beloruss. Gos. Univ. Ser. I** (1977), 22-23,93; MR 57:9853.

HAGER, ANTHONY W.

[85] Algebraic closures of ℓ-groups of continuous functions, **Rings of Continuous Functions (Cincinnati, Ohio: 1982)**, Marcel Dekker, New York, 1985.

(See also Aron and Dashiell.)

HAGER, ANTHONY W.; MADDEN, JAMES J.

[83] Majorizing-injectivity in abelian lattice-ordered groups, **Rend. Sem. Mat. Univ. Padova** 69 (1983), 181-194; MR 85h:06039.

HAGER, ANTHONY W.; ROBERTSON, LEWIS

[79] On the embedding of a ring in an Archimedean ℓ-group, **Canad. J. Math.** 31 (1979), 1-8; MR 80:16031.

HAHN, H.

[07] Über die nichtarchimedischen Größen-systeme, **Sitz. ber. K. Akad. der Wiss., Math. Nat. Kl.** IIa 116 (1907), 601-655.

HAUSNER, M.; WENDEL, J.G.

[52] Ordered vector spaces, **Proc. Amer. Math. Soc.** 3 (1952), 977-982; MR 14:566.

HEINZER, WILLIAM

[69] Some remarks on complete integral closure, **J. Austral. Math. Soc.** 9 (1969), 310-314; MR 40:254.

HENDRIKSEN, MELVIN; ISBELL, J.R.

[62] Lattice-ordered rings and function rings, **Pacific Math. J.** 12 (1962), 533-565.

HICKMAN, J.L.

[76] Groups of automorphisms of linearly ordered sets, **Bull. Austral. Math. Soc.** 15 (1976), 13-32; MR 55:204. Corrigenda: **ibid.**, 15 (1976), 317-318; MR 57:6156.

HILL, PAUL

[72] On the complete integral closure of a domain, **Proc. Amer. Math. Soc.** 36 (1972), 26-30; MR 46:7225.

[73] On free abelian ℓ-groups, **Proc. Amer. Math. Soc.** 38 (1973), 53-58; MR 47:1714.

HILL, PAUL; MOTT, JOE L.

[73] Embedding theorems and generalized discrete ordered abelian groups, **Trans. Amer. Math. Soc.** 175 (1973), 283-297; MR 47:102.

HION, YA. V.

[54] Archimedean ordered rings (in Russian), **Uspehi Mat. Nauk.**, (N.S.) 9, no.4 (1954),237-242; MR 16:442.

HÖLDER, O.

[01] Die Axiome der Quantität und die Lehre vom Maß, **Ber. Verh. Sächs. Ges. Wiss. Leipzig, Math.-Phys. Cl.** (1901), 1-64.

HOLLAND, W. CHARLES

[60] A totally ordered integral domain with a convex left ideal which is not an ideal, **Proc. Amer. Math. Soc.** 11 (1960), 703; MR 23:A2446.

[63a] The lattice-ordered group of automorphisms of an ordered set, **Michigan Math. J.** 10 (1963), 399-408; MR 28:1237.

[63b] Extensions of ordered groups and sequence completions, **Trans. Amer. Math. Soc.** 108 (1963), 71-80; MR 26:3795.

[65a] The interval topology of a certain ℓ-group, **Czech. Math. J.** 15(90) (1965), 311-314; MR 31:1308.

[65b] A class of simple lattice-ordered groups, **Proc. Amer. Math. Soc.** 16 (1965), 326-329; MR 30:3927.

[65c] Transitive lattice-ordered permutation groups, **Math. Z.** 87 (1965), 420-433; MR 31:2310.

[69] The characterization of generalized wreath products, **J. Algebra** 13 (1969), 152-172; MR 41:1884.

[74a] Ordered permutation groups, **Permutations** (Actes Colloq., Univ. Rene Descartes, Paris, 1972), 57-64, Gauthier-Villars, Paris, 1974; MR 49:8914.

[74b] Outer automorphisms of ordered permutation groups, **Proc. Edinburgh Math. Soc.** (2)19 (1974/75), 331-344; MR 52:7995.

[76a] The largest proper variety of lattice ordered groups, **Proc. Amer. Math. Soc.** 57 (1976), 25-28; MR 10688.

[76b] Equitable partitions of the continuum, **Fund. Math.** 92 (1976), 131-133; MR 54:5076.

[77] Group equations which hold in lattice-ordered groups, **Symposia Math.** 21 (1977), 365-378; MR 56:11676, which refers to Zbl 374:06011.

[79] Varieties of ℓ-groups are torsion classes, **Czech. Math. J.** 29(104), 11-12; MR 80b:06017.

[80] Trying to recognize the real line, 131-134 in [BKS].

[83] A survey of varieties of lattice-ordered groups, **Universal algebra and lattice theory (Puebla, 1982)**, 153-158; Lecture Notes in Math., 1004, Springer, Berlin, 1983; MR 84j:06020.

[84a] Classification of lattice-ordered groups, 151-155 in **Orders: description and roles**, North Holland, Amsterdam, 1984.

[84b] Intrinsic metrics for lattice-ordered groups, **Algebra Universalis** 19 (1984), 142-150; MR 85k:06012.

[85a] Varieties of automorphism groups of orders, **Trans. Amer. Math. Soc.** 288 (1985), 755-763; MR 86c:06022.

[85b] Intrinsic metrics for lattice-ordered groups, 99-106 in [PT].

[85c] A note on lattice orderability of groups, **Algebra Universalis** 20 (1985), 130-131.

(See also Conrad, Glass and Gurevich.)

HOLLAND, W. CHARLES; MARTINEZ, JORGE

[79] Accessibility of torsion classes, **Algebra Universalis** 9 (1979), 199-206; MR 80g:06021.

HOLLAND, W. CHARLES; MEKLER, ALAN H.; REILLY, NORMAN

[86] Varieties of lattice-ordered groups in which prime powers commute, to appear.

HOLLAND, W. CHARLES; MEKLER, ALAN H.; SHELAH, SAHARON

[85] Lawless order, **Order** 1 (1985), 383-397; MR 86m:06032.

HOLLAND, W. CHARLES; MCCLEARY, STEPHEN H.

[69] Wreath products of ordered permutation groups, **Pacific J. Math.** 31 (1969), 703-716; MR 41:3350.

[79] Solvability of the word problem in free lattice-ordered groups, **Houston J. Math.** 5 (1979), 99-105; MR 80f:06018.

HOLLAND, W. CHARLES; REILLY, NORMAN

[86a] Structures and laws of the Scrimger varieties of lattice-ordered groups, to appear.

[86b] Metabelian varieties of ℓ-groups which contain no non-abelian o-groups, to appear.

HOLLAND, W. CHARLES; SCRIMGER, E.

[72] Free products of lattice ordered groups, **Algebra Universalis** 2 (1972), 247-254; MR 47:8384.

HOLLEY, FRIEDA KOSTER

[73] The ideal and new interval topologies on ℓ-groups, **Fund. Math.** 79 (1973), 187-197; MR 49:172.

HOLLISTER, HERBERT A.

[65] **Contributions to the theory of partially ordered groups**, Thesis, Univ. of Michigan, 1965.

[78] Nilpotent ℓ-groups are representable, **Algebra Universalis** 8 (1978), 65-71; MR 57:5857.

HUDAĬBERDIEV, V.N.

[74] Topological ℓ-groups (in Russian), **Dokl. Akad. Nauk UzSSR** (1974), 6-7; MR 51:5440.

HUIJSMANS, C.B.

[76] Riesz spaces for which every ideal is a projection band, **Nederl. Akad. Wetensch. Proc. Ser. A = Indag. Math.** 38 (1976), 30-35; MR 53:7897.

HUSS, MARY E.

[84a] **Varieties of lattice-ordered groups**, Ph.D. dissertation, Simon Fraser University, 1984.

[84b] The lattice of lattice-ordered subgroups of a lattice-ordered group, **Houston J. Math.** 10 (1984), 503-505; MR 86h:06033.

HUSS, MARY E.; REILLY, NORMAN R.

[84] On reversing the order of a lattice-ordered group, **J. Algebra** (91)(1984), 176-191; MR 86i:06023.

[85] On reversing the order of a lattice-ordered group, 107-110 in [PT].

ISBELL, J.R.

[65] A structure space for certain lattice-ordered groups and rings, **J. London Math. Soc.** 40 (1965), 63-71; MR30:3155.

(See also Henriksen.)

ISLAMOV, A.N.

[73] Finite dimensionality of locally bicompact lattice-ordered groups (in Russian), **Dokl. Akad. Nauk UzSSR** (1973), 3-4; MR 48:5948.

[75a] The structure of commutative compactly generated lattice-ordered topological groups (in Russian), **Dokl. Akad. Nauk UzSSR** (1975), 3-4; MR 53:7895.

[75b] Structure of bicompactly generated commutative topological lattice-ordered groups, and self-generating topological lattice- ordered groups (in Russian), **Voronež. Gos. Univ. Trudy Naučn. Issled. Inst. Mat. VGU** Vyp. 20 (1975), 71-73; MR 57:9630.

[76] Self-generating topological lattice-ordered groups (in Russian), **Izv. Akad. Nauk UzSSR Ser. Fiz.-Mat. Nauk** (1976), 19-22,86; MR 55:7880.

ISLAMOV, A.N.; MIRONOV, A.V.

[72] On finitely based topological lattice-ordered groups (Russian), **Dokl. Akad. Nauk. UzSSR** (1972), 3-5; MR 47:5165.

IWASAWA, KENKICHI

[43] On the structure of conditionally complete lattice-groups, **Jap. J. Math.** 18 (1943), 777-789; MR 7:373.

IWATA, KAZUO

[70] On Ky Fan's theorem and its application to the free vector lattice, **Mem. Muroran Inst. Tech.** 7 (1970), 315-328; MR 43:133.

JAFFARD, PAUL

[50a] Applications de la théorie des filets, **C.R. Acad. Sci. Paris** 230 (1950), 1125-1126; MR 11:579.

[50b] Theorie des filets dans les groupes reticules, **C.R. Acad. Sci. Paris** 230 (1950), 1024-1025; MR 11:579.

[50c] Nouvelles applications de la théorie des filets, **C.R. Acad. Sci. Paris** 230 (1950), 1631-1632; MR 11:640.

[50d] Groupes archimédiens et para-archimédiens, **C.R. Acad. Sci. Paris** 231 (1950), 1278-1280; MR 12:480.

[53] Contribution à l'etude des groupes ordonnés, **J. Math. Pure Appl.** 32 (1953), 203-280; MR 15:284.

[54] Extension des groupes réticulés et applications, **Publ. Sci. Univ. Alger. Ser A** 1 (1954), 197-222; MR 17:346.

[56a] Extensions des groupes ordonnés, **Séminaire A. Chatelet et P. Dubreil de la Faculte des Sciences de Paris**, 1953/1954. Algèbre et theorie des nombres. 2e tirage multigraphie, pp. 11-01-11-10. Secretariat mathematique, Paris, 1956; MR 18:465.

[56b] Sur la théorie algébrique de la croissance, **C.R. Acad. Sci. Paris** 243 (1956), 1383-1385; MR 18:465.

[56c] Réalisation des groupes complètement réticulés, **Bull. Soc. Math. France** 84 (1956), 295-305; MR 18:790.

[57] Sur les groupes réticulés associes a un groupe ordonne, **Publ. Sc. Univ. Alger., Ser A** 2 (1957), 173-203; MR 19:13.

[61] Solution d'un problème de Krull, **Bull. Sci. Math.** 85 (1961), 127-135; MR 27:5828.

JAKUBÍK, JÁN

[59a] Konvexe Ketten in ℓ-Gruppen, **Časopis Pěst. Mat.** 84 (1959), 53-63; MR 21:3493.

[59b] On a class of lattice-ordered groups (Russian), **Časopis Pěst. Mat.** 84 (1959), 150-161; MR 21:7254.

[59c] On the principal ideals of a lattice-ordered group (Russian), **Czech. Math. J.** 9 (1959), 528-543; MR 22:12144.

[60] On a property of lattice-ordered groups, **Časopis Pěst. Mat.** 85 (1960), 51-59; MR 22:5667.

[61] Über eine Klasse von ℓ-Gruppen, **Acta Fac. Nat. Univ. Comenianae** 6 (1961), 267-273; MR 28:2156.

[62a] Über Teilbunde der ℓ-Gruppen, **Acta Sci. Math. Szeged** 23 (1962), 249-254; MR 26:2375.

[62b] The interval topology of an ℓ-group, **Mat. Fyz. Časopis Sloven. Akad. Vied** 12 (1962), 209-211; MR 29:174.

[63a] Über ein Problem von Paul Jaffard, **Arch. Math.** 14 (1963), 16-21; MR 32:4195.

[63b] Representation and extension of ℓ-groups, **Czech. Math. J.** 13(88) (1963), 267-283; MR 30:2091.

[64] Über Verbandsgruppen mit zwei Erzeugenden, **Czech. Math. J.** 14(89) (1964), 444-454; MR 30:1196.

[65a] Kompakt erzeugte Verbandsgruppen, **Math. Nachr.** 30 (1965), 193-201; MR 33:7439.

[65b] Über die Intervalltopologie auf einer halbgeordneten Gruppe, **Mat.-Fyz. Časopis Sloven. Akad. Vied** 15 (1965), 257-272; MR 34:4387.

[66] Die Dedekindschen Schnitte im direkten Produkt von halbgeordneten Gruppen, **Mat.-Fyz. Časopis Sloven. Akad. Vied** 16 (1966), 329-336; MR 35:1524.

[68] Higher degrees of distributivity in lattices and lattice-ordered groups, **Czech. Math. J.** 18(93) (1968), 356-376; MR 37:1283.

[69] On some problems concerning disjointness in lattice-ordered groups, **Acta Fac. Rerum Natur. Univ. Comenian. Math. Publ.** 22 (1969), 47-56; MR 42:162.

[70a] ℓ-subgroups of a lattice-ordered group, **J. London Math. Soc.** (2)2 (1970), 366-368; MR 41:1618.

[70b] On subgroups of a pseudo lattice-ordered group, **Pacific J. Math.** 34 (1970), 109-115; MR 42:5876.

[72a] Cantor-Bernstein theorem for lattice-ordered groups, **Czech. Math. J.** 22(97) (1972), 159-175; MR 45:6718.

[72b] Cardinal properties of lattice ordered groups, **Fund. Math.** 74 (1972), 85-98; MR 46:1672.

[72c] Distributivity in lattice-ordered groups, **Czech. Math. J.** 22(97) (1972), 108-125; MR 48:3834.

[72d] Homogeneous lattice-ordered groups, **Czech. Math. J.** 22(97) (1972), 325-337; MR 47:3273.

[73a] On σ-complete lattice ordered groups, **Czech. Math. J.** 23(98) (1973), 164-174; MR 47:6581.

[73b] Lattice ordered groups of finite breadth, **Colloq. Math.** 27 (1973), 13-20; MR 49:173.

[74a] Splitting property of lattice ordered groups, **Czech. Math. J.** 24(99) (1974), 257-269; MR 50:2016.

[74b] Normal prime filters of a lattice ordered group, **Czech. Math. J.** 24(99) (1974), 91-96; MR 50:204.

[74c] Lattice ordered groups with complete epimorphic images, **Colloq. Math.** 31 (1974), 21-28; MR 51:304.

[75a] Cardinal sums of linearly ordered groups, **Czech. Math. J.** 25(100) (1975), 568-575; MR 52:7996.

[75b] Conditionally orthogonally complete ℓ-groups, **Math. Nachr.** 65 (1975), 153-162; MR 51:305.

[75c] Products of torsion classes of lattice ordered groups, **Czech. Math. J.** 25(100) (1975), 576-585; MR 52:7997.

[76a] Strongly projectable lattice ordered groups, **Czech. Math. J.** 26(101) (1976), 642-652; MR 55:2701.

[76b] Lattice ordered groups with cyclic linearly ordered subgroups, **Časopis Pěst. Mat.** 101 (1976), 88-90; MR 55:12594.

[76c] Principal projection bands of a Riesz space, **Colloq. Math.** 36 (1976), 195-203; MR 55:3739.

[77] Radical mappings and radical classes of lattice ordered groups, **Symposia Math.**, 21 (1977), 451-477.

[78a] Archimedean kernel of a lattice ordered group, **Czech. Math. J.** 28(103) (1978), 140-154; MR 57:3033.

[78b] Orthogonal hull of a strongly projectable lattice ordered group, **Czech. Math. J.** 28(103) (1978), 484-504; MR 58:21891.

[78c] Generalized Dedekind completion of a lattice-ordered group, **Czech. Math. J** 28(103), 294-311; MR 58:27685.

[78d] Maximal Dedekind completion of an abelian lattice-ordered group, **Czech. Math. J.** 28(103), 611-631; MR 81j:06022.

[79] On algebraic operations of a lattice-ordered group, **Colloq. Math.** 41 (1979/80), 35-44; MR 81c:06020.

[80a] Products of radical classes of lattice ordered groups, **Acta. Math. Univ. Comenian** 39 (1980), 31-42; MR 82h:06024.

[80b] Weak isomorphisms of abelian lattice ordered groups, **Czech. Math. J.** 30(105) (1980), 438-444; MR 84g:06031.

[80c] Generalized lattice identities in lattice-ordered groups, **Czech. Math. J.** 30(105) (1980), 127-134; MR 81d:06024.

[80d] Isometries of lattice-ordered groups, **Czech. Math. J.** 30(105) (1980), 142-152; MR 81d:06025.

[81a] On isometries of nonabelian lattice-ordered groups, **Math. Slovaca** 31 (1981), 171-175; MR 82e:06016.

[81b] On value selectors and torsion classes of lattice-ordered groups, **Czech. Math. J.** 31(106) (1981), 306-313; MR 82m:06019a.

[81c] Prime selectors and torsion classes of lattice-ordered groups, **Czech. Math. J.** 31(106) (1981), 325-337; MR 82m:06019b.

[81d] On the lattice of torsion classes of lattice ordered groups, **Czech. Math. J.** 31(106) (1981), 510-513; MR 83h:06020.

[81e] On the lattice of radical classes of linearly ordered groups, **Studia Sci. Math. Hung.** 16 (181), 77-86; MR 86:06010.

[82a] On linearly ordered subgroups of a lattice ordered group, **Časopis Pěst. Mat.** 107 (1982), 175-179,188; MR 83g:06022.

[82b] Distributivity of intervals of torsion radicals, **Czech. Math. J.** 32(107) (1982), 548-555; MR 84f:06025.

[82c] Torsion radicals of lattice ordered groups, **Czech. Math. J.** 32(107) (1982), 347-363; MR 84:06021.

[82d] Projectable kernel of a lattice-ordered group, **Universal algebra and applications (Warsaw, 1978)**, 105-112, Banach Center Publ., Warsaw, 1982; MR 86c:06023.

[83a] On K-radical classes of lattice ordered groups, **Czech. Math. J.** 33(108) (1983), 149-163; MR 84f:06026.

[83b] On lexico extensions of lattice ordered groups, **Math. Slovaca** 33 (1983), 81-84; MR 84g:06027.

[83c] On the lattice of semisimple classes of linearly ordered groups, **Časopis Pěst. Mat.** 108 (1983), 183-190; MR 86:06011a.

[84] Kernels of lattice ordered groups defined by properties of sequences, **Časopis Pěst. Mat.** 109 (1984), 290-298.

[86] Radical subgroups of lattice-ordered groups, **Czech. Math. J.**, 36 (111) (1986), 285-297.

(See also Dreveňák.)

JAKUBÍK, JÁN; KOLIBIAR, MILAN

[83] Isometries of multilattice groups, **Czech. Math. J.** 33(108) (1983), 602-612; MR 85e:06023.

JAKUBÍKOVÁ, MÁRIA

[62] On some subgroups of ℓ-groups(in Russian), **Mat.-Fyz. Časopis Sloven. Akad. Vied** 12 (1962), 97-107; MR 29:173.

[71] Konvexe gerichtete Untergruppen der Rieszchen Gruppen, **Mat. Časopis Sloven. Akad. Vied** 21 (1971), 3-8; MR 46:1673.

[73a] Abgeschlossene vollstandige ℓ-Untergruppen der Verbandsgruppen, **Mat. Časopis Sloven. Akad. Vied** 23 (1973), 55-63; MR 48:2019.

[73b] Über die B-Potenz einer teilweise geordneten Gruppe, **Mat. Časopis Sloven. Akad. Vied** 23 (1973), 231-239; MR 49:2487.

[74] The nonexistence of free complete vector lattices, **Časopis Pěst. Mat.** 99 (1974), 142-146; MR 50:12854.

[75] Distributivitat des ℓ-Untergruppenverbandes einer Verbandsgruppe, **Mat. Časopis Sloven. Akad. Vied** 25 (1975), 189-192; MR 52:13559.

[78a] Totally inhomogeneous lattice ordered groups, **Czech. Math. J.** 28(103), 594-610; MR 58:16455.

[78b] On complete lattice ordered groups with two generators. I., **Math. Slovaca** 28 (1978), 389-406; MR 81e:06027a.

[79] On complete lattice ordered groups with two generators. II., **Math. Slovaca** 29 (1979), 271-287; MR 81e:06027b.

[82] Completions of lattice ordered groups, **Math. Slovaca** 32 (1982), 127-141; MR 83m:06023.

[83] Hereditary radical classes of linearly ordered groups, **Časopis Pěst. Mat.**108 (1983), 199-207; MR 86:06011b.

JAMBU-GIRAUDET, M.

[83] Bi-interpretable groups and lattices, **Trans. Amer. Math. Soc.** 278 (1983), 253-269; MR 84g:06028.

(See also Glass.)

JOHNSON, D.G.; KIST, J.E.

[62] Prime ideals in vector lattices, **Canad. J. Math.** 14 (1962), 517-528; MR 25:2010.

JOHNSON, D.G.; MACK, J.E.

[67] The Dedekind completion of C(X), **Pacific J. Math.** 20 (1967), 231-243; MR 35:2150.

KALMAN, J.A.

[56] An identity for ℓ-groups, **Proc. Amer. Math. Soc.** 7 (1956), 931-932; MR 18:274.

[60] Triangle inequality in ℓ-groups, **Proc. Amer. Math. Soc.** 11 (1960), 395; MR 22:1626.

KAMINSKIĬ, T.E.

[64] Direct sums and lexicographical extensions of lattice ordered groups (in Russian), **Moskov. Oblast. Ped. Inst. Učen. Zap.** 150 (1964), 179-200; MR 34:258.

[73] A new proof of the Jaffard-Conrad theorem (in Russian), **Groups and modules; game theory**, 26-31. Moskov. Oblast. Ped. Inst., Moscow, 1973; MR 52:13560.

KAPPOS, D.A.; KEHAYOPULU, NIOVI

[71] Some remarks on the representation of lattice ordered groups, **Math. Balkanica** 1 (1971), 142-143; MR 44:5263.

KEIMEL, KLAUS

[70] **Représentation de groupes et d'anneaux réticulés par des sections dans des faisceaux**, Thesis, Faculte des Sciences, Universite de Paris, Paris, 1970; MR 43:3194.

[71] The representation of lattice-ordered groups and rings by sections in sheaves, **Lectures on the applications of sheaves to ring theory (Tulane Univ. Ring and Operator Theory Year, 1970-1971, Vol. III)**, 1-98, Lecture Notes in Math., Vol. 248, Springer, Berlin, 1971; MR 54:10099.

[72] A unified theory of minimal prime ideals, **Acta Math. Sci. Hungar.** 23 (1972), 51-69; MR 47:6586.

(See also Bigard.)

KEISLER, MICHAEL

[77] A type of nearest point set in a complete ℓ-group , **Proc. Amer. Math. Soc.** 67 (1977), 189-197; MR 57:3034.

KENNY, OTIS

[75a] **Lattice-ordered Groups**, Ph.D. Dissertation, University of Kansas, 1975.

[75b] The completion of an abelian ℓ-group , **Canad. J. Math.** 27 (1975), 980-985; MR 52:13562.

[78] The Archimedean kernel of a lattice-ordered group, **Publ. Math. Debrecen** 25 (1978), 53-60; MR 58:10654.

(See also Anderson.)

KENOYER, DAVID

[84] Recognizability in the lattice of convex ℓ-subgroups of a lattice-ordered group, **Czech. Math. J.** 34 (109) (1984), 411-416; MR 86e:06017.

KHISAMIEV, N. G.

[66] Universal theory of lattice-ordered abelian groups, **Algebra i Logika** 5 (1966), 71-76; MR 34:2727.

KHISAMIEV, N.G.; KOKORIN, A.I.

[66] An elementary classification of lattice-ordered abelian groups with a finite number of fibers (in Russian), **Algebra i Logika** 5 (1966), 41-50; MR 33:7433.

KHUON, DAVITH

[70a] Cardinal des groupes réticulés: Complete archimédien d'un groupe réticulé, **C.R. Acad. Sci. Paris Ser. A-B** 270 (1970), A1150-A1153; MR 42:1735.

[70b] Groupes réticulés doublement transitifs, **C.R. Acad. Sci. Paris Ser. A-B** 270 (1970), A708-A709; MR 41:6746.

KOKORIN, A.I.

[62] On a class of lattice-ordered groups (in Russian), **Ural. Gos. Univ. Mat. Zap.** 3 (1962), 37-38; MR 29:5936.

[63] Methods for the lattice-ordering of a free Abelian group with a finite number of generators (in Russian), **Ural. Gos. Univ. Mat. Zap.** 4 (1963), 45-48; MR 29:5937.

(See also Khisamiev.)

KOKORIN, A.I.; KOZLOV, G.T.

[68] Extensively enlarged elementary theory of lattice-ordered groups having a finite number of filets (in Russian) **Algebra i Logika** 7 (1968), 91-103; MR 38:4298.

KOLDUNOV, A.V.

[79] A construction of the o-completion of Archimedean ℓ-group s with order identity (in Russian), **Dokl. Akad. Nauk UzSSR** (1979), 10-11; MR 80m:06017.

[80] Conditions for coincidence of the K-completion of an Archimedean ℓ-group with its o-completion (in Russian), **Modern algebra**, 50-57, Leningrad Gos. Ped. Inst., Leningrad, 1980; MR 81m:06041.

[87] Singularities of the structure of o-completion of an Archimedian ℓ-group with strong identity (Russian), **Czech. Math. J.** 37(112) (1987), 1-6.

KODUNOV, A.V.; ROTKOVIČ, G. JA.

[80] Archimedean lattice-ordered groups with the splitting property, 43-45 in [BKS]; MR 82a:06031.

[87] Archimedean lattice-ordered groups with the splitting property (Russian), **Czech. Math. J.** 37(112) (1987), 7-18.

KONTOROVIČ, P.G.

[64] Questions on fully ordered and lattice-ordered groups, **Spisy Prir. Fak. Univ. Brne** 9 (1964), 472-473.

KONTOROVIČ, P.G.; KUTYEV, K.M.

[59] On the theory of lattice-ordered groups (in Russian), **Izv. Vysš. Učenbnich. Zaved. Mat.** 10 (1959), 112-120; MR 24:A175.

KOPYTOV, V.M.

[75] Lattice-ordered locally nilpotent groups (in Russian), **Algebra i Logika** 14 (1975), 407-413; MR 53:5410.

[79] Free lattice-ordered groups (English translation), **Algebra and Logic** 18 (1979), 259-270; MR 81i:06018.

[81] Ordered groups (in Russian), **Algebra. Topology. Geometry.** Vol 19, 3-29,276, Akad. Nauk SSSR , Vsesoyuz. Inst. Nauchn. i Techn. Informatsii, Moscow, 1981; MR 83a:06024.

[82] Nilpotent lattice-ordered groups, **Sibirsk. Mat. Zh.** 23 (1982), 127-131,224; MR 84b:06016.

[83] Free lattice-ordered groups (in Russian), **Sibirsk. Mat. Zh.** 24 (1983), 120-124,192; MR 84d:06024.

[85] Nonabelian varieties of lattice-ordered groups in which every solvable ℓ-group is abelian (Russian), **Mat. Sb.** 126(168) (1985), 247-266, 287; MR 86j:06010.

(See also Gurchenkov)

KOPYTOV, V.M.; MEDVEDEV, N.I.

[77] Varieties of lattice-ordered groups (English translation), **Algebra and Logic** 16 (1977), 281-285; MR 58:27686.

[85] Varieties of lattice-ordered groups (Russian), **Uspekhi Mat. Nauk**, 40 (1985), 117-128, 199.

KRULL, W.

[32] Allgemeine Bewertungstheorie, **J. Reine Angew. Math.** 167 (1932), 160-196.

KUNDU, S. K.

[72] Measures in a Boolean algebra with values in a lattice-ordered group, **Math. Student** 40 (1972), 31-310; MR 52:14230.

KUNDU, S. K.; LAHIRI, B. K.

[79] On lattice-ordered group-valued measures, **Math. Student** 45 (1979), 1-9.

KUTYEV, K.M.

[56] On regular lattice-ordered groups (in Russian), **Uspechi Mat. Nauk** 11 (1956), 256.

[58] On the theory of lattice-ordered groups (in Russian), **Uspechi Mat. Nauk** 13 (1958), 238-239.

[68] SL-isomorphism of a lattice-ordered group (in Russian), **Dokl. Akad. Nauk SSSR** 179 (1968), 775-778; MR 37:6223.

[70] PS-isomorphism of a lattice-ordered group (in Russian), **Mat. Zametki** 7 (1970), 537-544; MR 41:8316.
(See also Konotorovič.)
LACAVA, FRANCESCO
[80] Observations on the theory of lattice-ordered abelian groups, **Boll. Un. Mat. Hal. A** 5 (1980), 319-322; MR 82:03032.
LANGFORD, E.S.
[65] Some results on linear operators on lattice groups, **Amer. Math. Monthly** 72 (1965), 841-846; MR 33:4163.
LANTZ, DAVID
[75] Finite Krull dimension, complete integral closure, and GCD domains, **Comm. Alg.** 3 (1975), 951-958; MR 52:3131.
LARSEN, M.; LEWIS, W.; SHORES, T.
[74] Elementary divisor rings and finitely presented modules, **Trans. Amer. Math. Soc.**, 187 (1974), 231-248; MR 49:280.
LEVI, F.W.
[13] Arithmetische Gesetze im Gebiet diskreter Gruppen, **Rend. Palermo** 35 (1913), 225-236.
[42] Ordered groups, **Proc. Indian Acad. Sci., Sect. A** 16 (1942), 256-263; MR 4:192.
LINÉS ESCARDÓ, E.; MALLOL BALMAÑA, R.
[52] On ℓ-groups (in Spanish), **Revista Mat. Hisp.-Amer.** (4)12 (1952), 129-136; MR 14:616.
LLOYD, JUSTIN T.
[64] **Lattice-ordered groups and o-permutation groups**, Ph.D. Dissertation, Tulane University, 1964.
[65] Representations of lattice-ordered groups having a basis, **Pacific J. Math.** 15 (1965), 1313-1317; MR 32:2489.
[67] Complete distributivity in certain infinite permutation groups, **Michigan Math. J.** 14 (1967), 393-400; MR 36:2544.
(See also Byrd.)
LOCI, ELIZABETH
[76] Compatible tight Riesz orders on C(X), **J. Austral. Math. Soc.** 22 (1976), 371-379; MR 55:232.
(See also Davis.)
LOONSTRA, F.
[51a] Discrete groups, **Nederl. Akad. Wetensch.** 54 = **Indag. Math.** 13 (1951), 162-168; MR 13:13.
[51b] The classes of partially ordered groups, **Compositio Math.** 9 (1951), 130-140; MR 13:625.
[77] Classes of partially ordered groups, **Symposia Math.** 21 (1977), 335-348; MR 58:430.
LORENZ, K.
[62] Über Strukturverbände von Verbandsgruppen, **Acta. Math. Acad. Sci. Hungar.** 13 (1962), 55-67; MR 25:2979.
LORENZEN, PAUL
[39] Abstrakte Begründung der multiplikativen Idealtheorie, **Math. Z.** 45 (1939), 533-553; MR 1:101.
[49a] Über halbgeordnete Gruppen, **Arch. Math.** 2 (1949), 66-70; MR 11:640.
[49b] Über halbgeordnete Gruppen, **Math. Z.** 52 (1949), 483-526; MR 11:497.
[53] Die Erweiterung halbgeordneter Gruppen zu Verbandsgruppen, **Math. Z.** 58 (1953), 15-24.
ŁOS, J.
[54] On the existence of linear order in a group, **Bull. Acad. Polon. Sci. Cl. III.** 2 (1954), 21-23; MR 16:564.
LOY, R.J.; MILLER, J.B.
[72] Tight Riesz groups, **J. Austral. Math. Soc.** 13 (1972), 224-240; MR 45:6721.
LUXEMBURG, W.A.J.; MOORE, L.C.
[67] Archimedean quotient Riesz spaces, **Duke Math. J.** 34 (1967), 725-739; MR 36:651.

MACNEILLE, H.

[37] Partially ordered sets, **Trans. Amer. Math. Soc.** 42 (1937), 416-460.

MADDEN, JAMES

[85] ℓ-groups of piecewise linear functions, 117-124 in [PT].

(See also Aron, Glass and Hager.)

MADELL, ROBERT L.

[69] Embeddings of topological lattice-ordered groups, **Trans. Amer. Math. Soc.** 146 (9169), 447-455; MR 40:4183.

[70] Chains which are coset spaces of $t\ell$-groups, **Proc. Amer. Math. Soc.** 25 (1970), 755-759; MR 41:8317.

[80] Complete distributivity and α-convergence, **Czech. Math. J.** 30(105) (1980), 296-301; MR 81d:06026.

MAEDA, FUMITOMO; OGASAWARA, TOZIRO

[42] Representation of vector lattices (Japanese), **J. Sci. Hirosima Univ. Ser. A** 12 (1942), 17-35; MR 10:544.

MALLOL BALMAÑA, R.

[52] Note (in Spanish), **Revista Mat. Hisp.Amer.** (4)12 (1952), 137; MR 14:616.

(See also Linés Escardó.)

MARTINEZ, JORGE

[71a] Approximation by archimedean lattice cones, **Pacific J. Math.** 36 (1971), 427-437; MR 43:3177.

[71b] Essential extensions of partial orders on groups, **Trans. Amer. Math. Soc.** 162 (1971), 35-61; MR 45:5053.

[72a] Hereditary properties and maximality conditions with respect to essential extensions of lattice group orders, **Trans. Amer. Math. Soc.** 166 (1972), 339-350; MR 45:5054.

[72b] Free products in varieties of lattice-ordered groups, **Czech. Math. J** 22(97) (1972), 535-553; MR 47:98.

[72c] Tensor products of partially ordered groups, **Pacific J. Math.** 41 (1972), 771-789; MR 47:3274.

[73a] A hom-functor for lattice-ordered groups, **Pacific J. Math.** 48 (1973), 169-183; MR 48:10939.

[73b] Archimedean-like classes of lattice-ordered groups, **Trans. Amer. Math. Soc.** 186 (1873), 33-49; MR 48:10940.

[73c] Structure of archimedean lattices, **Proceedings of the University of Houston Lattice Theory Conference (Houston, Tex. 1973)**, 295-305, Dept. Math., Univ. Houston, Houston, Tex., 1973; MR 52:13537.

[73d] Archimedean lattices, **Algebra Universalis** 3 (1973), 247-260; MR 50:1996.

[73e] Free products of abelian ℓ-groups, **Czech. Math. J.** 23(98) (1973), 349-361; MR 51:7987.

[74a] Varieties of lattice-ordered groups, **Math. Z.** 137 (1974), 265-284; MR 50:6961.

[74b] The hyper-archimedean kernel sequence of a lattice-ordered group, **Bull. Austral. Math. Soc.** 10 (1974), 337-349; MR 50:2017.

[75a] Doubling chains, singular elements and hyper-z-ℓ-groups, **Pacific J. Math.** 61 (1975), 502-506; MR 53:5411.

[75b] Torsion theory for lattice-ordered groups, **Czech. Math. J.** 25(100) (1975), 284-299; MR 52:10536.

[76] Torsion theory for lattice-ordered groups. II. Homogeneous ℓ-groups, **Czech. Math. J** 26(101) (1976), 93-100; MR 52:10537.

[77a] Pairwise splitting lattice-ordered groups, **Czech. Math. J.** 27(102) (1977), 545-551; MR 56:15526.

[77b] Is the lattice of torsion classes algebraic?, **Proc. Amer. Math. Soc.** 63 (1977), 9-14; MR 58:27687.

[79] Nilpotent lattice-ordered groups, **Algebra Universalis** 9 (1979), 329-338; MR 81a:06021.

[80a] Lexicographic centers of lattice-ordered group, 29-42 in [BKS]; MR 82b:06017.

[80b] The fundamental theorem on torsion classes of lattice-ordered groups, **Trans. Amer. Math. Soc.** 259 (1980), 311-317; MR 81i:06019.

[80c] General torsion theory for lattice-ordered groups, U. of Florida Lecture Notes, 1980.

[81] Prime selectors in lattice-ordered groups, **Czech. Math. J** 31(106) (1981), 206-217; MR 82h:06025.

[82] Almost finite-valued ℓ-groups, **Rend. Sem. Mat. Univ. Padova** 67 (1982), 75-84; MR 84g:06029.

[85] Abstract ideal theory, 125-138 in [PT].

(See also Anderson, Baird and Holland.)

MASTERSON, J.J.

[68] Structure spaces of a vector lattice and its Dedekind completion, **Neder. Akad. Wetensch. Proc.** 71 (1968), 468-478; MR 43:872.

MATSUSHITA, SHIN-ICHI

[51] On the foundation of orders in groups, **J. Inst. Polytech. Osaka City Univ. Ser. A Math.** 2 (1951), 19-22; MR 13:624.

MCALISTER, DONALD B.

[65] On multilattice groups, **Proc. Cambridge Philos. Soc.** 61 (1965), 621-638; MR 31:95.

[66] On multilattice groups II, **Proc. Cambridge Philos. Soc.** 62 (1966), 149-164; MR 32:7654.

[71] Multilattice groups with a lexico basis, **Proc. Roy. Irish Acad. Sect. A** 71 (1971), 53-72; MR 45:8594.

(See also Conrad.)

MCCLEARY, STEPHEN H.

[69] The closed prime subgroups of certain ordered permutation groups, **Pacific J. Math.** 31 (1969), 745-753; MR 42:1736.

[70] Generalized wreath products viewed as sets with valuation, **J. Algebra** 16 (1970), 163-182; MR 42:380.

[72a] Pointwise suprema of order-preserving permutations, **Illinois J. Math.** 16 (1972), 69-75; MR 45:3275.

[72b] Closed subgroups of lattice-ordered permutations groups, **Trans. Amer. Math. Soc.** 173 (9172), 303-314; MR 47:97.

[72c] o-primitive ordered permutation groups, **Pacific J. Math.** 40 (1972), 349-372; MR 47:1710.

[73a] o-2 transitive ordered permutation groups, **Pacific J. Math.** 49 (1973), 425-429; MR 50:2018.

[73b] o-primitive ordered permutation groups. II., **Pacific J. Math.** 49 (1973), 431-443; MR 50:2019.

[73c] The lattice-ordered group of automorphisms of an α-set, **Pacific J. Math.** 49 (1973), 417-424; MR 50:6962.

[76] The structure of intransitive ordered permutation groups, **Algebra Universalis** 6 (1976), 229-255; MR 54:12597.

[80] A solution of the word problem in free normal-valued lattice-ordered groups, 107-129 in [BKS]; MR 82h:06026.

[81] The lateral completion of an arbitrary lattice-ordered group (Bernau's proof revisited), **Algebra Universalis** 13 (1981), 251-263; MR 82k:06022.

[82] The word problem in free normal valued lattice-ordered groups: a solution and practical shortcuts, **Algebra Universalis** 14 (1982), 317-348; MR 83f:06027.

[85a] Free lattice-ordered groups represented as o-2 transitive ℓ-permutation groups, **Trans. Amer. Math. Soc.** 290 (1985), 69-79; MR 86m:06034a.

[85b] An even better representation for free lattice-ordered groups, **Trans. Amer. Math. Soc.** 290 (1985), 81-100; MR 86m:06034b.

[85c] Free lattice-ordered groups, 139-154 in [PT].

[86] Lattice-ordered groups whose lattices of convex ℓ-subgroups guarantee noncommutativity, **Order** 3 (1986), 307-315.

(See also Arora, Davis, Glass and Holland.)

MCNULTY, GEORGE F.

[80] Classes which generate the variety of all lattice-ordered groups, 135-140 in [BKS].

MEDVEDEV, N. YA.
[77] Lattices of varieties of lattice-ordered groups and Lie groups, **Algebra i Logika** 16 (1977), 40-45, 123; MR 58:16456.
[80] Decomposition of free ℓ-groups into an ℓ-direct product (in Russian), **Sibirsk. Mat. Z.** 21 (1980), 63-69,190; MR 82b:06018.
[82a] ℓ-varieties without an independent basis of identities (in Russian), **Math. Slovaca** 32 (1982), 417-425; MR 84b:06017.
[82b] On the theory of varieties of lattice-ordered groups (in Russian), **Czech. Math. J** 32(107) (1982), 364-372; MR 84c:06020.
[83a] The lattice of radicals of finitely generated ℓ-groups, **Math. Slovaca** 33 (1983), 185-188; MR 84h:06017.
[83b] Coverings in a lattice of ℓ-varieties (Russian), **Algebra i Logika** 22 (1983), 53-60; MR 86f:08009.
[83c] Certain questions of the theory of partially ordered groups, **Algebra i Logika** 22 (1983), 435-442; MR 85k:06113.
[84a] Lattice of o-approximable ℓ-varieties (Russian), **Czech. Math. J.** 34(109), 6-17; MR 85h:06036.
[84b] Free products of ℓ-groups (Russian), **Algebra i Logika** 23 (1984), 493-511,599; MR 87b:06032.
[85] Quasivarieties of ℓ-groups and groups (Russian), **Sibirsk. Math. Z.** 26 (1985), 111-117, 206.
(See also Kopytov.)
MICHIURA, TADASCHI
[49a] On a definition of lattice ordered groups, **J. Osaka Inst. Sci. Tech. Part I.** 1 (1949), 27; MR 11:497.
[49b] On a definition of lattice-ordered groups. II., **J. Osaka Inst. Sci. Tech. Part I.**, 1 (1949), 117-119; MR 12:389.
[50] Sur les groupes semi-ordonnés, **C.R. Acad. Sci. Paris** 231 (1950), 1403-1404; MR 12:480.
[52a] Sur les groupes semi-ordonnés. II., **C.R. Acad. Sci. Paris** 234 (1952), 1422-1423; MR 14:19.
[52b] Sur les groupes semi-ordonnés. III., **C.R. Acad. Sci. Paris** 234 (1952), 1521-1522; MR 14:19.
MILLER, JOHN BORIS
[73a] Quotient groups and realization of tight Riesz groups, **J. Austral. Math. Soc.** 16 (1973), 416-430; MR 49:4898.
[73b] A characterization of weak projectability, **Bull. Austral. Math. Soc.** 8 (1973), 205-209; MR 47:3275.
[76] Subdirect representation of tight Riesz groups by hybrid products, **J. Reine Angew. Math.** 283/84 (1976), 110-124; MR 54:204.
[78a] The order-dual of a TRL group. I., **J. Austral. Math. Soc.** 25 (1978), 129-141; MR 80f:06019.
[78b] Direct summands in ℓ-groups, **Proc. Roy. Soc. Edinburgh Sect. A** 81 (1978), 175-186; MR 80b:06018.
(See also Loy.)
MIRONOV, A.V.
[75] The structure of nilpotent lattice-ordered groups of bicompact descent (in Russian), **Dokl. Akad. Nauk UzSSR** (1975), 7-9; MR 53:234.
[76a] The structure of nilpotent topological lattice ordered groups of bicompact origin. Basic results (in Russian), **Dokl. Akad. Nauk SSSR** 228 (1976), 300-302; MR 54:202.
[76b] Compactly generated nilpotent lattice-ordered groups (in Russian), **Taskent. Gos. Univ. Naucn. Trudy Vyp. 490 Voprosy Matematiki** (1976), 123-128, 266; MR 57:12330.
(See also Antonovksii and Islamov.)
MOČKOŘ, JIŘ'I
[78] Topological groups of divisibility, **Colloq. Math.** 39 (1978), 301-311; MR 80i:06016.

MOORE, L.C., JR.

[70] The lifting property in Archimedean Riesz spaces, **Nederl. Akad. Wetensch. Proc. Ser. A** 73 = **Indag. Math.** 32 (1970), 141-150; MR 41:3353.

(See also Luxemburg.)

MOTT, JOE L.

[73] The group of divisibility and its applications, **Conference on Commutative Algebra (Lawrence, Kansas, 1972)**, 194-208, Lecture Notes in Mathematics 311, Springer, Berlin, 1973.

[75] Generalized discrete ℓ-groups, **J. Austral. Math. Soc.** 20 (1975), 281-289; MR 52:5514.

MUNDICI, DANIELE

[87] Every abelian ℓ-group with two positive generators is ultrasimplicial, **J. of Algebra** 105 (1987), 236-241.

MURATA, KENTARO

[77] On lattice ideals of arithmetical lattice-ordered groups (in Japanese), **Sugaku** 29 (1977), 75-77; MR58:432.

NAKANO, HIDEGORO

[40] Teilweise geordnete Algebra, **Proc. Imp. Acad. Tokyo** 16 (1940), 437-441; MR 2:343.

[41a] Teilweise geordnete Algebra, **Jap. J. Math.** 17 (1941), 425-511; MR 3:210.

[41b] Eine Spektraltheorie, **Proc. Phys.-Math. Soc. Japan** (3)23 (1941), 485-511; MR 3:210.

[41c] Über normierte teilweisegeordnete Moduln, **Proc. Imp. Acad. Tokyo** 17 (1941), 311-317; MR 7:249.

[41d] Über die Charakterisierung des allgemeinen C-Raumes, **Proc. Imp. Acad. Tokyo** 17 (1941), 301-307; MR 7:249.

[41e] Über das System aller Stetigen Funktiones auf einem topologischen Raum, **Proc. Imp. Acad. Tokyo** 17 (1941), 308-310.

[42a] Über ein lineares Funktional auf dem teilweise geordneten Modul, **Proc. Imp. Acad. Tokyo** 18 (1942), 548-552; MR 7:249.

[42b] Über die Charakterisierung des allgemeinen C-Raumes. II., **Proc. Imp. Acad. Tokyo** 18 (1942), 280-286; MR 7:249.

[43] Über die Stetigkeit des normierten teilweise geordneten Moduls, **Proc. Imp. Acad. Tokyo** 19 (1943), 10-11; MR 7:249.

[48] On the product of relative spectra, **Ann. Math.** (2) 49 (1948), 281-315; MR 9:445.

NAKAYAMA, TADASI

[42a] On Krull's conjecture concerning completely integrally closed integrity domains. I., **Proc. Imp. Acad. Tokyo** 18 (1942), 185-187; MR 7:236.

[42b] On Krull's conjecture concerning completely integrally closed integrity domains. II., **Proc. Imp. Acad. Tokyo** 18 (1942), 233-236; MR 7:236.

[42c] Note on lattice-ordered groups, **Proc. Imp. Acad. Tokyo** 18 (1942), 1-4; MR 7:240.

NAKAYAMA, TADASI; YOSIDA, KOSAKU

[42] On the semiordered ring and its application to the spectral theorem, **Proc. Imp. Acad. Tokyo** 18 (1942), 555-560; MR 7:253.

[43] On the semiordered ring and its application to the spectral theorem. II., **Proc. Imp. Acad. Tokyo** 19 (1943), 144-147; MR 7:253.

NEUMANN, B. H.

[37] Identical relations in groups. I, **Math. Ann.** 114 (1937), 506-525.

[49] On ordered groups, **Amer. J. Math.** 71 (1949), 1-18.

[60] Embedding theorems for ordered groups, **J. London Math. Soc.** 35 (1960), 503-512; MR 24:A176.

NEUMANN, K.

[68] Über darstellungen von Verbanden mit o-Idealen gerichteter Gruppen, **Arch. Math. (Brno)** 4 (1968), 61-73; MR 41:1612.

OGASAWARA, TOZIRO

[42a] Compact metric Boolean algebras and vector lattices, **J. Sci. Hirosima Univ. Ser. A** 11 (1942), 125-128; MR 10:46.

[42b] Theory of vector lattices. I., **J. Sci. Hirosima Univ. Ser. A** 12 (1942), 37-100; MR 10:545.

[43a] On Frechet lattices. I. (in Japanese), **J. Sci. Hirosima Univ. Ser. A** 12 (1943), 235-248; MR 10:544.

[43b] Commutativity of Archimedean semiordered groups, **J. Sci. Hirosima Univ. Ser. A** 12 (1943), 249-254; MR 10:544.

[43c] Remarks on representation of vector lattices (Japanese), **J. Sci. Hirosima Univ. Ser. A** 12 (1943), 217-234; MR 10:544.

[44a] Remarks on a vector lattice with a metric function, **J. Sci. Hirosima Univ. Ser. A** 13 (1944), 317-325; MR 10:544.

[44b] Theory of vector lattices. II. (Japanese), **J. Sci. Hirosima Univ. Ser. A** 13 (1944), 41-161; MR 10:545.

(See also Maeda.)

OHKUMA, T.

[55] Sur quelques ensembles ordonnés linéairement, **Fund. Math.** 43 (1955), 326-337; MR 18:868.

OHM, JACK

[66] Some counterexamples related to integral closure of $D[[X]]$, **Trans. Amer. Math. Soc.** 122 (1966), 321-333.

[69] Semi-valuations and groups of divisibility, **Canad. J. Math.** 21 (1969), 576-591.

OHNISHI, MASAO

[50] On linearization of ordered groups, **Osaka Math. J.** 2 (1950), 161-164; MR 13:436.

ORIHARA, MASAE

[42] On the regular vector lattice, **Proc. Imp. Acad. Tokyo** 18 (1942), 525-529; MR 7:250.

PAPADOPOULOU, SUSAN

[68] Remarks on the completion of commutative lattice-groups with respect to order convergence, **Bull. Soc. Math. Grece (N.S.)** 9 (1968), 138-142; MR 40:5519.

PAPANGELOU, FREDOS

[64] Order convergence and topological completion of commutative lattice-groups, **Math. Ann.** 155 (1964), 81-107; MR 30:4699.

[65] Some considerations on convergence in abelian lattice-groups, **Pacific J. Math.** 15 (1965), 1347-1364; MR 32:7655.

PAPERT, D.

[62] A representation theory for lattice-groups, **Proc. London Math. Soc.** (3) 12 (1962), 100-120.

PEDERSEN, F.D.

[69] A representation for a class of lattice ordered groups, **Trans. Amer. Math. Soc.** 140 (1969), 117-126; MR 39:1377.

[70] Implications between conditions on ℓ-groups, **Canad. J. Math.** 22 (1970), 1-6; MR 41:128.

[74] Spitz in ℓ-groups, **Czech. Math. J** 24(99) (1974), 254-256; MR 50:2020.

[75] Epimorphisms in the category of ℓ-groups, **Proc. Amer. Math. Soc.** 53 (1975), 311-317.

PETROVA, N.L.

[80] Polars and rectifying subgroups in lattice-ordered groups (Bulgarian), **Godishnik Vissh. Uchebn. Zaved. Prilozhna Mat.** 16 (1980), 67-72 (1981); MR 83e:06027.

PIERCE, KEITH R.

[72a] Amalgamations of lattice ordered groups, **Trans. Amer. Math. Soc.** 172 (1972), 249-260; MR 48:3835.

[72b] Amalgamating Abelian ordered groups, **Pacific J. Math.** 43 (1972), 711-723.

[73] The structure of a lattice-ordered group as determined by its prime subgroups, **Proc. Amer. Math. Soc.** 40 (1973), 407-412; MR 49:2488.

[76] Amalgamated sums of abelian ℓ-groups, **Pacific J. Math.** 65 (1976), 167-173; MR 55:2702.

[85] Embedding theorems for lattice-ordered groups, 163-170 in [PT].

(See also Glass.)

PINSKER, A.

[49] Extended semiordered groups and spaces, **Učen. Zapiski Leningrad Gos. Ped. Inst.** 86 (1949), 236-265.

PLOTKIN, B. I.

[69] On the semigroup of radical classes of groups, **Sibirsk. Math. Z.** 10 (1969), 991-1008; MR 41:323.

PONS, MONTSERRAT

[82] T-topologies on a lattice ordered group, **Stochastica** 6 (1982), 25-37; MR 84k:06023.

POWELL, WAYNE B.

[81] Projectives in a class of lattice ordered modules, **Algebra Universalis** 13 (1981), 24-40; MR 83a:06026.

[82] Boolean hypolattices with applications to ℓ-groups, **Arch. Math. (Basel)** 39 (1982), 535-540.

[83] Injectives in a class of lattice ordered modules, **Houston J. Math.** 9 (1983), 275-287; MR 85b:06014.

[85] On isometries in abelian lattice ordered groups, **J. Indian Math. Soc. (N.S.)** 46 (1982), 189-194 (1985).

POWELL, WAYNE B.; TSINAKIS, CONSTANTINE

[82] Free products of abelian ℓ-groups are cardinally indecomposable, **Proc. Amer. Math. Soc.** 86 (1982), 385-390; MR 84a:06011.

[83a] Free products in the class of abelian ℓ-groups, **Pacific J. Math.** 104 (1983), 429-442; MR 84c:06022.

[83b] The distributive lattice free product as a sublattice of the abelian ℓ-group free product, **J. Austral. Math. Soc.** 34 (1983), 92-100; MR 84g:06032.

[84] Free products of lattice-ordered groups, **Algebra Universalis** 18 (1984), 178-198; MR 85m:06034.

[85a] Amalgamation of lattice-ordered groups, 171-178 in [PT].

[85b] Meet irreducible varieties of lattice-ordered groups, **Algebra Universalis** 20 (1985), 262-263; MR 87a:06036.

[86] Disjointness conditions for free products, **Arch. Math.** 46 (1986), 491-498.

[87] Covers of the abelian variety of lattice-ordered groups, preprint.

PSHENICHNOV, V.I.

[78a] The structure of lattice-ordered groups with a topology (Russian), **Dokl. Akad. Nauk UzSSR** (1978), 9-11; MR 80:06018.

[78b] The construction of certain lattice-ordered groups with a topology (Russian), **Taskent. Gos. Univ. Sb. Naučn. Trudov No. 573 Funktional. Anal.** (1978), 74-76, 111; MR 81:06028.

[78c] Finite dimensionality of locally bicompact lattice-ordered groups with a relatively closed n-continuous order (Russian), **Taskent. Gos. Univ. Sb. Naučn. Trudov No. 573 Funktional. Anal.** (1978), 72-74, 111; MR 81d:06027.

[80] A topological analogue of a theorem of Conrad for ℓ-groups with a basis (Russian), **Dokl. Akad. Nauk UzSSR** (1980), 6-8; MR 82c:06033.

[81a] On the structure of locally compact ℓ-groupsin the o-topology (in Russian), **Dokl. Akad. Nauk UzSSR** (1981), 3-5; MR 83e:06030.

[81b] On the structure of an ℓ-group with the o-topology and the o*-topology (in Russian), **Dokl. Akad. Nauk UzSSR** (1981), 5-7; MR 83j:06023.

RACHUNEK, JIRE

[73a] Lattice-ordered groups with minimal prime subgroups satisfying a certain condition, **Arch. Math. (Brno)** 9 (1973), 147-149; MR 50:9740.

[73b] Directed convex subgroups of ordered groups, **Sb. Praci Prirodoved. Fak. Univ. Palackeho v Olomouci** 41 (1973), 39-46; MR 50:9741.

[74] Prime subgroups of ordered groups, **Czech. Math. J** 24(99) (1974), 541-551; MR 50:9742.

[76] On extensions of orders of groups and rings, **Acta Math. Acad. Sci. Hungar.** 28 (1976), 37-40; MR 54:7351.

[78] Riesz groups with a finite number of disjoint elements, **Czech. Math. J** 28(103), 102-107; MR 57:12329.

[79] Z-subgroups of ordered groups, **Math. Slovaca** 29 (1979), 39-41; MR 81c:06021.
[82] A characterization of ordered groups by means of segments, **Sb. Praci Prirodoved. Fak. Univ. Palackeho v Olomouci Mat.** 21 (1982), 33-35; MR 84g:06030.
[84] Isometries in ordered groups, **Czech. Math. J** 34 (109)(1984), 334-341; MR 86a:06026.

RADNEV, P.
[70] A certain class of lattice-ordered multioperator groups (Bulgarian), **Vysš Ped. Inst. Plovdiv Naučn. Trud.** 8 (1970), 35-38; MR 45:6720.

RAMA BALA, D.V.; SASTRY, K.P.R.
[77] Additive set functions with values in a complete lattice ordered group, **Math. Student** 45 (1977), 7-12 (1979); MR 82a:06030.

RANGA RAO, P.
[78] Metric spaces with distances in a lattice-ordered group, **Math. Sem. Notes Kobe Univ.** 6 (1978), 189-200; MR 80:52017.

READ, JOHN A.
[73] ℓ-sous-groupes compressibles du groupe-réticulé A(S), **Séminaire P. Dubreil (25e annee: 1971/72)**, Algèbre, Fasc. i, Exp. No. 6, Secretariat Mathematique, Paris, 1973; MR 52:10538.
[74] Wreath products of nonoverlapping lattice ordered groups, **Canad. Math. Bull.** 17 (1974/75), 713-722; MR 52:5515.

REDFIELD, R.H.
[74] A topology for a lattice-ordered group, **Trans. Amer. Math. Soc.** 187 (1974), 103-125; MR 48:5949.
[75] Bases in completely distributive lattice-ordered groups, **Michigan Math. J.** 22 (1975), 301-307; MR 53:5413.
[76] Archimedean and basic elements in completely distributive lattice-ordered groups, **Pacific J. Math.** 63 (1976), 247-253; MR 54:203.

REILLY, NORMAN R.
[69] Some applications of wreath products and ultraproducts in the theory of lattice ordered groups, **Duke Math. J.** 36 (1969), 825-834; MR 40:4184.
[72] Permutational products of lattice ordered groups, **J. Austral. Math. Soc.** 13 (1972), 25-34; MR 45:3278.
[73] Compatible tight Riesz orders and prime subgroups, **Glasgow Math. J.** 14 (1973), 145-160; MR 49:174.
[75] Representations of ordered groups with compatible tight Riesz orders, **J. Austral. Math. Soc.** 20 (1975), 307-322; MR 52:224.
[81] A subsemilattice of the lattice of varieties of lattice ordered groups, **Canad. J. Math.** 33 (1981), 1309-1318; MR 83g:06023.
[83] Nilpotent, weakly abelian and Hamiltonian lattice-ordered groups, **Czech. Math. J** 33(108), 348-353; MR 85m: 06035.
[86] Varieties of lattice-ordered groups that contain no non-abelian o-groups are solvable, **Order** 3 (1986), 287-297.
(See also Holland and Huss.)

REILLY, NORMAN R.; WROBLEWSKI, ROGER
[81] Suprema of classes of generalized Scrimger varieties of lattice-ordered groups, **Math. Z.** 176 (1981), 293-309; MR 82k:06023.

REMA, P.S.
[65] On a class of topologies in lattice ordered groups, **J. Madras Univ.** B 35-36 (1965/66), 19-26; MR 42:2994.

RIBENBOIM, PAULO
[58a] Sur quelques constructions de groupes réticulés et ℓ-équivalence logique entre l'different de filtres et d'ordres, **Summa Brasil. Math.** 4 (1958), 65-89; MR 21:5685.
[58b] Sur les groupes totalement ordonnés et l'arithmétique des anneaux de valuation, **Summa Brasil. Math.** 4 (1958), 1-64; MR 21:6396.
[60] Un théorème de réalisation des groupes réticulés, **Pacific J. Math.** 10 (1960), 305-308; MR 22:1624.

RICE, N.M.

[68] Multiplication in vector lattices, **Canad. J. Math.** 20 (1968), 1136-1149; MR 37:6732.

RIECAN, BELOSLAV

[76] On the lattice group valued measures, **Casopis Pest. Math.** 101 (1976), 343-349; MR 58:17036.

RIECAN, BELOSLAV; VOLAUF, PETER

[84] A technical lemma in lattice-ordered groups, **Acta Math. Univ. Comenian** 44(45) (1984), 31-36.

RIESZ, FREDERIC

[40] Sur quelques notions fondamentales dans la theorie generale des operations lineaires, **Ann. Math.** 41 (1940), 174-206; MR 1:147.

ROTKOVIČ, G. JA.

[73a] The normality of the realization multiplication in partially ordered groups (Russian), **Optimizacija Vyp.** 12(29) (1973), 100-104, 155; MR 53:13071.

[73b] Relatively uniformly complete semiordered groups (Russian), **Functional analysis**, No. 2, 183-194, Ul'janovsk Gos. Univ., Ulyanovsk, 1973; MR 58:21892.

[75] On σ-complete lattice-ordered groups (Russian), **Czech. Math. J.** 25(100) (1975), 279-281; MR 51:5441.

[77a] Disjunctively complete Archimedean partially ordered groups (Russian), **Czech. Math. J.** 27(102), 523-527; MR 57:3036.

[77b] Some forms of completeness in Archimedean lattice-ordered groups (Russian), **Functional analysis**, No. 9: Harmonic analysis on groups, 138-143, Ul'janovsk. Gos. Ped. Inst., Ulyanovsk, 1977; MR 57:12332.

[80] Lattice-ordered groups with cyclic linearly ordered subgroups (in Russian), **Modern Algebra**, Leningrad. Gos. Ped. Inst., Leningrad, 1980, 119-123; MR 82f:06027.

(See also Kodunov.)

SARACINO, DAN; WOOD, CAROL

[83] Finitely generic abelian lattice-ordered groups, **Trans. Amer. Math. Soc.** 277 (1983), 113-123.

[84] An example in the model theory of abelian lattice-ordered groups, **Algebra Universalis** 19 (1984), 34-37; MR 85k:06014.

(See also Glass.)

SCHMIDT, KLAUS D.

[85] A common abstraction of Boolean rings and latticed-ordered groups, **Compositio Math.** 51 (1985), 51-62.

SCOTT, THOMAS J.

[74] Monotonic permutations of chains, **Pacific J. Math.** 55 (1974), 583-594; MR 52:5516.

SCRIMGER, E.B.

[75] A large class of small varieties of lattice-ordered groups, **Proc. Amer. Math. Soc.** 51 (1975), 301-306; MR 52:5517.

(See also Holland.)

SHELDON. B.

[73] Two counterexamples involving complete integral closure in finite dimensional Prüfer domains, **J. Algebra** 27 (1973), 462-474; MR 48: 11097.

SHERMAN, B.F.

[74] Cauchy completion of partially ordered groups, **J. Austral. Math. Soc.** 18 (1974), 222-229; MR 57:12333.

[75] Cauchy completion of abelian tight Riesz groups, **J. Austral. Math. Soc.** 19 (1975), 62-73; MR 53:5417.

SHIPOSH, JÁN

[77] Integrals on lattice-ordered groups, **Math. Slovaca** 27 (1977), 431-439; MR 80f:28016.

ŠIK, FRANTIŠEK

[56] Zur Theorie der halbgeordneten Gruppen (Russian), **Czech. Math. J.** 6(81) (1956), 1-25; MR 18:464.

[58a] Automorphismen geordneter Mengen, **Casopis Pest. Mat.** 83 (1958), 1-22; MR 20:2290.

[58b] Über Summen einfach geordneter Gruppen, **Czech. Math. J.** 8(83) (1958), 22-53; MR 20:5810.

[62] Kompakt erzeugte vollstandige ℓ-Gruppen, **Bul. Inst.Politehn. Iasi** (N.S.) 8(12) (1962), 5-8; MR 32:2496.

[63] Structure and realizations of lattice-ordered groups (Spanish), **Mem. Fac. Ci. Univ. Habana Ser. Mat.** 1 (1963/64), 1-29; MR 32:145.

[65a] Types speciaux de realixations des groupes reticules, **C.R. Acad. Sci. Paris** 261 (1965), 4948-4949; MR 32:7656.

[65b] Sous-groupes simples et idéaux simples des groupes réticulés, **C.R. Acad. Sci. Paris** 261 (1965), 2791-2793; MR 33:2738.

[66] Struktur und Realisierungen von Verbandsgruppen. III. Einfache Untergruppen und einfache Ideale, **Mem. Fac. Ci. Univ. Habana Ser. Math.** 1 (1966), 1-20; MR 34:7681.

[67] Struktur und Realisierungen von VerbandsGruppen. V.
Schwache Einheiten in Verbandsgruppen, **Math. Nachr.** 33 (1967), 221-229; MR 42:3002.

[68a] Archimedische kompakt erzeugte Verbandsgruppen, **Math. Nachr.** 38 (1968), 323-340; MR 39:4072.

[68b] Struktur under Realisierung von Verbandsgruppen. IV. Spezielle Typen von Realisierungen, **Mem. Fac. Ci. Univ. HabanaSer. Mat.** 1 (1968), 19-44; MR 48:5950.

[70] Completely lattice preorderable groups (in Russian), **Moskov. Oblast. Ped. Inst. Ucen. Zap.** 282 (1970), 146-150; MR 45:3279.

[71] Verbandsgruppen, deren Komponentenverband kompakt erzeugt ist, **Arch. Math. (Brno)** 7 (1971), 101-121; MR 47:101.

[73] Closed and open sets in topologies induced by lattice ordered vector groups, **Czech. Math. J** 23(98) (1973), 139-150; MR 47:8387.

[81] Topology on regulators of lattice ordered groups. I. Topology induced by an ℓ-group, **Math. Slovaca** 31 (1981), 417-428; MR 83c:06018.

[82a] Γ-regulator and Π-regulator of a lattice ordered group, **Math. Slovaca** 32 (1982), 105-116; MR 83k:06018.

[82b] Topology on regulators of lattice ordered groups. II. Completely regular regulators, **Math. Slovaca** 32 (1982), 35-48; MR 84c:06021.

[82c] Regulators of type α of lattice ordered groups, **Math. Slovaca** 32 (1982), 209-227; MR 84d:06025.

ŠIMBIREVA, E.P.

[72] Fundamental trends in the development of the theory of partially ordered groups (Russian), **Moskov. Oblast. Ped. Inst. Učen. Zap. 311 Metodika Mat., Element. Mat. Vysš. Algebra, Prikl. Mat. Vyp.** 9 (1972), 211-238; MR 58:16454.

ŠIRŠOVA, E.E.

[73] Pseudo-lattice-ordered groups (Russian), **Groups and modules; game theory**, 10-18, Moskov. Oblast. Ped. Inst., Moscow, 1973; MR 52:13561.

ŠMARDA, BOHUMIL

[67] Topologies in ℓ-groups, **Arch. Math. (Brno)** 3 (1967), 69-81; MR 36:6331.

[69] Some types of topological ℓ-groups, **Spisy Prirod Fak. Univ. Brno** 507 (1969), 341-351; MR 42:7821.

[70] Algebraically and topologically compact ordered groups, **Spisy Přírod. Fak. Univ. Brno** (1970), 471-479; MR 45:138.

[76a] The lattice of topologies of topological ℓ-groups, **Czech. Math. J.** 26(101) (1976), 128-136; MR 53:5414.

[76b] Connectivity in tℓ-groups, **Arch. Math. (Brno)** 12 (1976), 1-7; MR 56:5387.

[85] Topologies corresponding to metrics on ℓ-groups, **Math. Slovaca** 35 (1985), 185-193.

SMIRNOV, D. M.

[67] On one-sided orders in groups with an ascending central series, **Algebra i Logika** 6 (1967), 77-87; MR 35:5372.

SMITH, JO E.

[76] **The lattice of ℓ-group varieties**, Ph.D. Dissertation, Bowling Green State U., 1976.

[80a] The lattice of ℓ-group varieties, **Trans. Amer. Math. Soc.** 257 (1980), 347-357; MR 81k:06026b.

[80b] ℓ-group varieties, 99-105 in [BKS]; MR 82c:08007.

[81] A new family of ℓ-group varieties, **Houston J. Math.** 7 (1981), 551-570; MR 83g:06024.

[84] Solvable and ℓ-solvable ℓ-groups, **Algebra Universalis** 18 (1984), 106-109.

SPEED, T.P.; STRZELECKI, E.

[71] A note on commutative ℓ-groups, **J. Austral. Math. Soc.** 12 (1971), 69-74; MR 43:3191.

STANKOVIČ, B.

[60] Completion d'un groupe réticulé, **Acad. Serbe Sci. Publ. Inst. Math.** 14 (1960), 115-122; MR 23:A3440.

STEINBERG, STUART

[80] Lattice-ordered modules of quotients, **J. Austral. Math. Soc.** 30 (1980/81), 243-251; MR 82e:06019.

STONE, M.H.

[40] A general theory of spectra. I., **Proc. Nat. Acad. Sci. U.S.A.** 26 (1940), 280-283; MR 1:338.

[41] A general theory of spectra. II., **Proc. Nat. Acad. Sci. U.S.A.** 27 (1941), 83-87; MR 2:318.

[49] Boundedness properties in function-lattices, **Canad. J. Math.** 1 (1949), 176-186; MR 10:546.

SWAMY, K.L.N.

[64] Autometrized lattice ordered groups. I., **Math. Ann.** 154 (1964), 406-412; MR 29:171.

[77] Isometries in autometrized lattice ordered groups. II., **Math. Sem. Notes. Kobe Univ.** 5 (1977), 211-214; MR 57:3037b.

[78] Isometries in autometrized lattice ordered groups, **Algebra Universalis** 8 (1978), 59-64; MR 57:3037a.

TEH, H.H.

[62] A note on ℓ-groups, **Proc. Edinburgh Math. Soc.** 13 (1962), 123-124; MR 26:56.

[67] Structure of commutative lattice ordered groups, **J. Nanyang Univ.** 1 (1967), 265-278; MR 38:4382.

TELLER, J. ROGER

[64] On the extensions of lattice-ordered groups, **Pacific J. Math.** 14 (1964), 709-718; MR 29:1269.

[65] On partially ordered groups satisfying the Riesz interpolation property, **Proc. Amer. Math. Soc.** 16 (1965), 1392-1400; MR 32:5749.

[68a] A theorem on Riesz groups, **Trans. Amer. Math. Soc.** 130 (1968), 254-264; MR 37:120.

[68b] On abelian pseudo lattice-ordered groups, **Pacific J. Math.** 27 (1968), 411-419; MR 38:3203.

(See also Byrd and Conrad.)

TODORINOV, S.A.

[82] Lattice orderable groups (Russian), **C.R. Acad. Bulgare Sci.** 35 (1982), 1189-1191; MR 84j:06022a.

TONG, DAO RONG

[82] The intrinsic topologies of a lattice group (Chinese), **J. China Univ. Sci. Tech.** (1982) Suppl. II, 1-9; MR 84k:06020.

TOPPING, DAVID M.

[65] Some homological pathology in vector lattices, **Canad. J. Math.** 17 (1968), 411-428; MR 30:4700.

TREVISAN, GIORGIO

[51] Sulla equivalenza archimedea relativa alle gruppo-strutture, **Rend. Sem. Mat. Univ. Padova** 20 (1951), 425-429; MR 14:19.

TSINAKIS, CONSTANTINE

[85] Projectable and strongly projectable lattice-ordered groups, **Algebra Universalis** 20 (1985), 57-76.

(See also Powell.)

VAIDA, D.

[60] On the isolated subgroups of a non-abelian ℓ-group (in Roumanian), **Com. Acad. R. P. Romine** 10 (1960), 935-939.

[62] Un probleme de G. Birkhoff, **C. R. Acad. Bulgare Sci.** 15 (1962), 801-803.

VAN METER, KARL M.

[71] Sur les groupes reticules et leur extension archimedienne, **Bull. Sci. Math.** (2)95 (1971), 59-64; MR 44:131.

[73] Les sous-groupes d'un groupe quasiréticulé, **Séminaire P. Dubreil (25e annee: 1971/72), Algèbre**, Fasc. 1, Secretariat Mathematique, Paris, 1973; MR 53:5412.

VEKSLER, A.I.

[69] A new construction of the Dedekind completion of vector lattices and divisible ℓ-groups(in Russian), **Siber. Math. J.** 10 (1969), 891-896; MR 41:5935.

[77] Archimedean free linear hulls of partially ordered groups (Russian), **Ordered sets and lattices, No. 4**, 11-23, 133, Izdat. Saratov. Univ., Saratov, 1977; MR 58:433.

(See also Geiler.)

VINOGRADOV, A.A.

[67] Ordered algebraic systems, **Algebra, Topology, Geometry, 1965** (Russian), 83-131, **Akad. Nauk SSSR Inst. Naucn. Tehn. Informacii**, Moscow, 1967; MR 35:6596.

[68] Ordered algebraic systems, **Algebra, Topology, Geometry, 1966** (Russian), 91-108, **Akad. Nauk SSSR Inst. Naucn. Tehn. Informacii**, Moscow, 1968; MR 38:4366.

[71] Nonaxiomatizability of lattice-orderable groups (Russian), **Sibirsk. Mat. Z.** 12 (1971), 463-464; MR 44:132.

VINOGRADOV, ANAT. A.; VINOGRADOV, ANDR. A.

[69] Nonlocalness of lattice-orderable groups, **Algebra i Logika** 8 (1969), 636-639; MR 43:1904.

VOVSI, S.M.

[72] On infinite products of classes of groups, **Sibirsk. Mat. Z.** 13 (1972), 272-285; MR 46:1916.

[73] On infinite products of hereditary classes of groups, **Latv. Mat. Yezegodnik** 13 (1973), 36-44; MR 49:421.

[83] On radical and coradical classes of ℓ-groups, Algebra Universalis 16 (1983), 159-162; MR 85c:06015.

VOLAUF, PETER

[77] Extension and regularity of ℓ-group-valued measures, **Math. Slovaca** 27 (1977), 47-53; MR 57:16534.

(See also Riecan.)

WAAIJERS, LEONARDUS J. M.

[68] **On the structure of lattice ordered groups**, Doctoral dissertation, Technical University of Delft, Uitgeverij Waltman, Delft, 1968; MR 38:3200.

WATERMAN, ALAN G.

[70] The normal completions of certain partially ordered vector spaces, **Proc. Amer. Math. Soc.** 25 (1970), 141-144; MR 41:129.

WEINBERG, ELLIOTT C.

[62] Completely distributive lattice-ordered groups, **Pacific J. Math.** 12 (1962), 1131-1137; MR 26:5064.

[63] Free lattice-ordered abelian groups, **Math. Ann.** 151 (1963), 187-199; MR 27:3720.

[65] Free lattice-ordered abelian groups. II., **Math. Ann.** 159 (1965), 217-222; MR 31:5895.

[66] o-projective ordered abelian groups, **J. Reine Angew. Math.** 224 (1966), 219-220; MR 34:1419.

[67] Embedding in a divisible lattice-ordered group, **J. London Math. Soc.** 42 (1967), 504-506; MR 36:91.

[77] Relative injectives, **Symposia Math.** 21 (1977), 555-564; MR 58:16457.

[80] Automorphism groups of minimal η_α-sets, 71-79 in [BKS]; MR 82c:06032.

[81] Real order-automorphism groups, **J. Austral. Math. Soc.** 31 (1981), 163-167; MR 83i:06020.

WILHELM, MAREK

[80] Completeness of ℓ-groups and of ℓ-seminorms, **Comment. Math. Prace Mat.** 21 (1980), 271-281; MR 81i:06023.

[84] Integral extension procedures in weakly α-complete lattice-ordered groups. I., **Studia Math.** 77 (1984), 423-435; MR 85j: 06016.

WIRTH, ANDREW

[72] Convergence in partially ordered groups, **Proc. Edinburgh Math. Soc.** (2)18 (1972/73), 239-246; MR 48:5953.

[73] Compatible tight Riesz orders, **J. Austral. Math. Soc.** 15 (1973), 105-111; MR 51:12653.

[74] Some Krein-Milman theorems for order-convexity, **J. Austral. Math. Soc.** 18 (1974), 257-261; MR 50:12853.

[75] Locally compact tight Riesz groups, **J. Austral. Math. Soc.** 19 (1975), 247-251; MR 51:7986.

WOLFENSTEIN, SAMUEL

[66] Sur les groupes réticulés archimédiennement complets, **C.R. Acad. Sci. Paris Ser. A-B** 262 (1966), A813-A816; MR 33:2739.

[67] Extensions archimédiennes non-commutatives de groupes réticulés commutatifs, **C.R. Acad. Sci. Paris Ser. A-B** 264 (1967), A1-A4; MR 34:5954.

[68a] Completes archimédiens de groupes réticulés à valeurs finies, **C.R. Acad. Sci. Paris Ser. A-B** 267 (1968), A592-A595; MR 38:3202.

[68b] Valeurs normales dans un groupe réticulé, **Atti Accad. Naz. Lincei Rend. Cl. Sci. Fis. Mat. Natur.** (8)44 (1968), 337-342; MR 38:3201.

[70a] Extensions archimédiennes de groupes réticulés transitifs, **Bull. Soc. Math. France** 98 (1970), 193-200; MR 42:5879.

[70b] **Contribution à l'étude des groupes réticulés: Extensions archimédiennes, Groupes à valeurs normales**, Dissertation, U. of Paris, 1970.

[73] Groupes réticulés singuliers, **Séminaire P. Dubreil (25e annee: 1971/72)**, Algèbre, Fasc. 1, Secretariat Mathematique, Paris, 1973; MR 53:5416.

[76] Représentation d'une classe de groupes Archimédiens, **J. Algebra** 42 (1976), 199-207; MR 54:5077.

[85] Semiprojectable ℓ-groups, **Czech. Math. J.** 35 (110) (1985), 385-390; MR 87b:06034.

(See also Bigard.)

WRIGHT, I.W.

[72] Divisibility of ordered groups, **Proc. Edinburgh Math. Soc.** (2)18 (1972/73), 81-83; MR 47:6583.

YOSIDA, KOSAKU

[41] On vector lattice with a unit, **Proc. Imp. Acad. Tokyo** 17 (1941), 121-124; MR 3:210.

[42] On the representation of the vector lattice, **Proc. Imp. Acad. Tokyo** 18 (1942), 339-342; MR 7:409.

(See also Fukamiya and Nakayama.)

ZAITZEVA, M.I.

[58] Right-ordered groups (in Russian), **Učen. Zap. Suisk. Gos. Ped. Inst.** 6 (1958), 205-226.

SUBJECT INDEX

AUTHOR INDEX

GPSR Compliance

The European Union's (EU) General Product Safety Regulation (GPSR) is a set of rules that requires consumer products to be safe and our obligations to ensure this.

If you have any concerns about our products, you can contact us on ProductSafety@springernature.com

In case Publisher is established outside the EU, the EU authorized representative is:

Springer Nature Customer Service Center GmbH
Europaplatz 3
69115 Heidelberg, Germany

Batch number: 10607224

Printed by Printforce, the Netherlands